D0021413

DATE DUE

*Q*THE EXPANDED
Quotable Einstein

*Q*THE EXPANDED
uotable Einstein

COLLECTED AND EDITED BY

Alice Calaprice

WITH A FOREWORD BY

Freeman Dyson

PRINCETON UNIVERSITY PRESS

PRINCETON AND OXFORD

Library of Congress Cataloging-in-Publication Data

Einstein, Albert, 1879–1955.
The expanded quotable Einstein / collected and edited by
Alice Calaprice ; with a foreword by Freeman Dyson.
    p. cm.
Includes bibliographical references and index.
ISBN 0-691-07021-0 (cl)
1. Einstein, Albert, 1879–1955—Quotations.
I. Calaprice, Alice.   II. Title.
QC16.E5 A25 2000
081—dc21   00-026873

This book has been composed in Palatino

The paper used in this publication meets the minimum requirements
of ANSI/NISO Z39.48-1992 (R1997) (*Permanence of Paper*)

www.pup.princeton.edu

Printed in the United States of America

10 9 8 7 6 5 4 3 2 1

FRONTISPIECE: Black and white drawing by New Jersey artist
Ben Shahn, 1953. (Courtesy of Susan Merians, Princeton)

In memory of my mother,

Rusan Abeghian

1911–1999

who was always
so proud of us all

Contents

The Quotations

Foreword

My excuse for writing this foreword is that I have
been for thirty years a friend and adviser to Princeton University Press, helping to smooth the way for
the huge and difficult project of publishing the Einstein Papers, a project in which Alice Calaprice is
playing a central role. After long delays and bitter
controversies, the publication project is now going
full steam ahead, producing a steady stream of volumes packed with scientific and historical treasures.

I knew Einstein only at second hand through his
secretary and keeper of the archives, Helen Dukas.
Helen was a warm and generous friend to grown-
ups and children alike. She was for many years our
children's favorite babysitter. She loved to tell stories about Einstein, always emphasizing his sense
of humor and his serene detachment from the passions that agitate lesser mortals. Our children remember her as a gentle and good-humored old
lady with a German accent. But she was also tough.
She fought like a tiger to keep out people who tried
to intrude upon Einstein's privacy while he was
alive, and she fought like a tiger to preserve the privacy of his more intimate papers after he died. She
and Otto Nathan were the executors of Einstein's
will, and they stood ready with lawsuits to punish

anyone who tried to publish Einstein documents without their approval. Underneath Helen's serene surface we could occasionally sense the hidden tensions. She would sometimes mutter darkly about unnamed people who were making her life miserable.

Einstein's will directed that the archive containing his papers should remain under the administration of Otto Nathan and Helen so long as they lived, and should thereafter belong permanently to the Hebrew University in Jerusalem. For twenty-six years after Einstein's death in 1955, the archive was housed in a long row of filing cabinets at the Institute for Advanced Study in Princeton. Helen worked every day at the archive, carrying on an enormous correspondence and discovering thousands of new documents to add to the collection.

In December 1981, Otto Nathan and Helen were both in apparently good health. Then, one night around Christmas, when most of the Institute members were on holiday, there was a sudden move. It was a dark and rainy night. A large truck stood in front of the Institute with a squad of well-armed Israeli soldiers standing guard. I happened to be passing by and waited to see what would happen. I was the only visible spectator, but I have little doubt that Helen was also present, probably supervising the operation from her window on the top floor of the Institute. In quick succession, a number of big wooden crates were brought down in the elevator from the top floor, carried out of the building through the open front door, and loaded onto the truck. The soldiers jumped on board and the truck drove away into the night. The next day, the archive was in its final resting place in Jerusalem. Helen continued to come to work at the Institute, taking care of her correspondence and tidying up the empty space where the archive had been. Six weeks later, suddenly and unexpectedly, she died. We never knew whether she had had a premonition of her death; in any case, she made sure that her beloved archive would be in safe hands before her departure.

After the Hebrew University took responsibility for the archive and after Otto Nathan's death in January 1987, the ghosts that had been haunting Helen quickly emerged into daylight. Robert Schulmann, a historian of science who had joined the Einstein Papers Project a few years earlier, received a tip from Switzerland that a secret cache of love letters, written around the turn of the century by Einstein and his first wife, Mileva Marić, might still exist. He began to suspect that the cache might be part of Mileva's literary estate, brought to California by her daughter-in-law Frieda, the first wife of Einstein's older son, Hans, after Mileva's death in Switzerland in 1948. Though Schulmann had received repeated assurances that the only extant letters were those dating from after Mileva's separation from Einstein in 1914, he was not convinced. He met in 1986 with Einstein's granddaughter, Evelyn, in Berkeley. Together they discovered a critical clue. Tucked away in an unpublished manuscript that Frieda had prepared about Mileva, but not part of the text, were notes referring with great immediacy to fifty-four love letters. The conclusion was obvious: these letters must be part of the group of more than four hundred in the hands of the Einstein Family Correspondence Trust, the legal entity representing Mileva's California heirs. Because Otto Nathan and Helen Dukas had earlier blocked publication of Frieda's biography, the Family Trust had denied them access to the correspondence and they had no direct knowledge of its contents. The discovery of Frieda's notes and the transfer of the literary estate to the Hebrew University afforded a new opportunity to pursue publication of the correspondence.

In spring 1986, John Stachel, at the time the editor responsible for the publication of the archive, and Reuven Yaron, of the Hebrew University, broke the logjam by negotiating a settlement with the Family Trust. Their aim was to have photocopies of the correspondence deposited with the publication project and with the Hebrew University. The crucial meeting took place in California, where Thomas Einstein, the physicist's oldest great-grandson and a trustee of the Family Trust, lives. The negotiators were disarmed when the young man arrived in tennis shorts, and a friendly settlement was quickly reached. As a result, the intimate letters became public. The letters to Mileva revealed Einstein as he really was, a man not immune from normal human passions and weaknesses. The letters are masterpieces of pungent prose, telling the sad old story of a failed marriage, beginning with tender and playful love, ending with harsh and cold withdrawal.

During the years when Helen ruled over the archive, she kept by her side a wooden box which she called her "Zettelkästchen"—her little box of snippets. Whenever in her daily work she came across an Einstein quote that she found striking or charming, she typed her own copy of it and put it in the box. When I visited her in her office, she would always show me the latest additions to the box. The contents of the box became the core of the book *Albert Einstein, the Human Side*, an anthology of Einstein quotes which she co-edited with Banesh Hoffmann and published in 1979. *The Human Side* depicts the Einstein that Helen wanted the world to see, the Einstein of legend, the friend of schoolchildren and impoverished students, the gently ironic philosopher, the Einstein without violent feelings and tragic mistakes. It is interesting to contrast the Einstein portrayed by Helen in *The Human Side* with the Einstein portrayed by Alice Calaprice in this book. Alice has chosen her quotes impartially from the old and the new documents. She does not emphasize the darker side of Einstein's personality, and she does not conceal it. In the brief section "On His Family," for example, the darker side is clearly revealed.

In writing a foreword to this collection, I am forced to confront the question whether I am committing an act of betrayal. It is clear that Helen would have vehemently opposed the publication of the intimate letters to Mileva and to Einstein's second wife, Elsa. She would probably have felt betrayed if she had seen my name attached to a book that contained many quotes from the letters that she abhorred. I was one of her close and trusted friends, and it is not easy for me to go against her express wishes. If I am betraying her, I do not do so lightheartedly. In the end, I salve my conscience with the thought that, in spite of her many virtues, she was profoundly wrong in trying to hide the true Einstein from the world. While she was alive, I never pretended to agree with her on this point. I did not try to change her mind, because her conception of her duty to Einstein was unchangeable, but I made it clear to her that I disliked the use of lawsuits to stop publication of Einstein documents. I had enormous love and respect for Helen as a person, but I never promised that I would support her policy of censorship. I hope and almost believe that, if Helen were now alive and could see with her own eyes that the universal admiration and respect for Einstein have not been diminished by the publication of his intimate letters, she would forgive me.

It is clear to me now that the publication of the intimate letters, even if it is a betrayal of Helen Dukas, is not a betrayal of Einstein. Einstein emerges from this collection of quotes, drawn from many different sources, as a complete and fully rounded human being, a greater and more astonishing figure than the tame philosopher portrayed in Helen's book. Knowledge of the darker side of Einstein's life makes his achievement in science and in public affairs even more miraculous. This book shows him as he was—not a superhuman genius but a human genius, and all the greater for being human.

A few years ago I had the good luck to be lecturing in Tokyo at the same time as the cosmologist Stephen Hawking. Walking the streets of Tokyo with Hawking in his wheelchair was an amazing experience. I felt as if I were taking a walk through Galilee with Jesus Christ. Everywhere we went, crowds of Japanese silently streamed after us, stretching out their hands to touch Hawking's wheelchair. Hawking enjoyed the spectacle with detached good humor. I was thinking of an account that I had read of Einstein's visit to Japan in 1922. The crowds had streamed after Einstein then as they streamed after Hawking seventy years later. The Japanese people worshiped Einstein as they now worshiped Hawking. They showed exquisite taste in their choice of heroes. Across the barriers of culture and language, they sensed a godlike quality in these two visitors from afar. Somehow they understood that Einstein and Hawking were not just great scientists but great human beings. This book helps to explain why.

Freeman Dyson

THE INSTITUTE FOR ADVANCED STUDY
PRINCETON, NEW JERSEY

Preface and Acknowledgments

> In the past it never occurred to me that every casual remark of mine would be snatched up and recorded. Otherwise I would have crept further into my shell.
>
> —*Einstein to his biographer Carl Seelig,*
> *October 25, 1953*

Albert Einstein was a prolific—and often thoughtful and gifted—writer, and he is immensely quotable. This I discovered when I began my work with the Einstein papers in 1978 preparing a computerized index of the duplicate Einstein archive, located at the time (along with the original archive) at the Institute for Advanced Study in Princeton. The job, under the direction of John Stachel, then the editor of *The Collected Papers of Albert Einstein*, required a perusal of all the documents—correspondence, writings, and third-party commentary. From these, our assistant Edith Laznovsky and I would glean certain bits of information and enter them into the not-so-user-friendly computer of the 1970s that was made available to us at the Princeton University cyclotron laboratory. I would often read these items—most of them in German—more thoroughly than necessary, simply because they were so engrossing.

I impulsively began to keep an index-card file of my favorite excerpts and quotations, and it is these cards that serve as the basis for this book.

Since I came to work at Princeton University Press and was appointed both in-house editor of the Press's huge publishing venture, *The Collected Papers of Albert Einstein*, and administrator of its accompanying translation project, I often received calls and letters from people asking for the source of some quotation or another, usually found on a calendar or heard on the radio and attributed to Einstein. At the same time, I had learned that the Einstein Papers Project editorial offices in Boston, the Firestone Library at Princeton University, and the library of the Institute for Advanced Study were also besieged with such inquiries. Most of the time we were not able—at least not easily or quickly—to establish the source or correct wording of the quotations. This situation, the blue plastic box of quotations on my shelf, and the interest of Trevor Lipscombe, the Press's physical sciences editor, gave me the idea for this book.

To come up with this selection, I have not only depended on my blue box but also searched through many other original sources plus Einstein biographies and additional secondary sources, as well as rechecked parts of the duplicate archive. I have not limited myself to quotations suitable for after-dinner speeches and epigraphs but have also included some less profound utterings that reflect various facets of Einstein's personality. Some of these statements may distress readers who have worshiped Einstein as a compassionate, tolerant, and flawless hero; see, for instance, his brusque reply to a Chilean official who requested some words of wisdom, his diary entry regarding the devout at the Wailing Wall in Jerusalem, and his ideas on women in science. Other readers may take pleasure in the fact that their worst prejudices against him, whether they be religious, philosophical, or political, are confirmed by his thoughts on abortion, marriage, communism, and world government. Still others will delight in his humor (see, for instance, the subsection on animals and pets under "Miscellaneous Subjects"), and will identify with him as he shares his thoughts on everything from youth to aging, from pipe smoking to going sockless.

But before rushing to judgment, one must take into consideration Einstein's age at the time of quotation and his milieu—the historical and cultural times in which he lived. Indeed, over his lifetime, he changed his mind or qualified his opinion on several topics—pacifism, the death penalty, and Zionism, for instance. In addition, although he used the now politically incorrect *mankind* and generic *he* when referring to people in general, professionally he did indeed dwell in a man's world. However, much of the use of "man" may be due to mistranslations of the German *Mensch*, which refers collectively to both men and women.

The organization of this book fell naturally into the categories listed alphabetically (after the sections "On Einstein Himself" and "On His Family") in the table of contents, and then into a larger "Miscellaneous" section, also organized alphabetically by subject. Within each category, the quotations are listed chronologically when I was able to establish the dates, and the undated ones in that category are lumped together at the end.

I quote from the original documents whenever possible. Among these are the Einstein Archive (I give the document numbers of the duplicate archives found in Princeton and Boston); the volumes in *The Collected Papers of Albert Einstein (CPAE)*; *Albert Einstein, the Human Side* by Helen Dukas and Banesh Hoffmann, which contains archival material selected by Einstein's longtime secretary, who was also his archivist; and the various books and journals in which particular articles first appeared. In addition, I often list reliable and easily available compilations such as *Ideas and Opinions*, so that readers can consult this more popular literature for complete text and context. (My page numbers refer to the editions cited in the Bibliography.) In the few instances where I could not find an original source, I relied on the secondary literature, such as biographies.

I have made every effort to verify references, but *The Quotable Einstein* cannot aspire to be a work of scholarship in the strictest sense—I cannot claim to have used the best or most authoritative version of a translation, for example, as these often differ from book to book. If I found no translation, I used my own or relied on that of a friend. For the expanded edition, I retranslated those quotations that I had found awkward.

Needless to say, there must be many worthy words that I did not come across and that are hiding somewhere among the over 40,000 documents in the archive, so this effort can by no means be considered a complete book of quotations. But I hope that, for now, I have been able to present and document the most important or interesting ones. As this will be an ongoing project, with enlarged editions published every few years, I invite the reader to send me any quotations, along with documentation, that I may have missed. We will include these in the future editions. If I have inadvertently misquoted Einstein or given a false source, please let me know that as well.

I came across a few quotations whose sources I could not find, yet I—or people who have called me with inquiries—have seen or heard them attributed to Einstein. I have put these toward the back of the book in a section entitled "Attributed to Einstein"; my hope is that readers can lead me to the proper documentation.

To help the reader or researcher locate items, I have compiled two indexes: the Index of Key Words will help readers find familiar quotations, and the Subject Index, prepared by Lys Ann Shore for the enlarged edition, will lead them to topics of particular interest.

Finally, I wish to record my acknowledgments to those who have helped in the preparation of this book. First, I thank the Hebrew University of Jerusalem for permission to include material from the Einstein archive; the Philosophical Library for permission to include material reprinted in *Ideas and Opinions*; and Crown Publishers for permission to use the quotations drawn from Jamie Sayen's *Einstein in America*.

I am grateful for the help, interest, and support of my family, friends, and colleagues. In particular, I would like to thank my colleagues at Princeton University Press who have shown enthusiasm for this project from the start, especially Trevor Lipscombe, Eric Rohmann, and Emily Wilkinson (now no longer at the Press). In addition, special thanks go to my longtime friend and the Press's former managing editor, Janet Stern, for showing me that even a professional editor's writing needs to be edited. Computer whiz Linda Moran patiently initiated me into the world of word processing, then showed great proficiency in composing the original edition of this book. Our senior designer, Jan Lilly, designed both the original edition and the current one with sensitivity and skill. Bing Lin Zhao of Boston University remained good-natured and uncommonly helpful when I repeatedly interrupted his work to enlist his help in computer searches, saving me hours of time. Evelyn Einstein graciously helped me update the family tree, and Mark Hazarabedian designed it with great care. My late mother, Rusan Abeghian, clipped Einstein material from newspapers in several languages.

I am also grateful to Freeman Dyson for taking time out from his busy schedule to write the foreword, even though he would have preferred seeing the original German in this volume as well. (A separate German edition is now available.) When I was looking through my old index cards, I came across one on which I had scribbled some remarks that Helen Dukas had made about him in 1978. Helen, who knew I was of Armenian descent on my mother's side, had told me about an article that Freeman, whom I had not yet met, had written for the *New Yorker* several years earlier about his visit to Armenia. After our discussion, she said something more about Freeman Dyson that is worthy of quotation in a book such as this: "He is a great man. My one regret is that he did not meet Professor Einstein. In the '50s, the professor mentioned that he had heard of this interesting young man. I told him I could arrange a meeting, but the professor said, 'Oh, no, I don't want to bother such an important young man!'" Unlike the polite Professor Einstein, I dared bother this man—to ask him to write a foreword for this book; and I am deeply grateful that he readily agreed.

Last but not least, Robert Schulmann, director of the Einstein Papers Project at Boston University, has, as always, been an invaluable friend and source of information and good cheer, even when I felt I was testing his patience. I hope that this book has met everyone's expectations.

PRINCETON, NEW JERSEY
JANUARY 1996 AND 2000

A Note about the Expanded Edition

Four years have passed since the publication of the original edition of *The Quotable Einstein*. These years have greatly enriched my life as I came to know what it's like to be on the other side of the publishing world—that is, what it's like to be author rather than editor, to be signing books in bookstores and for friends, to be reviewed and interviewed.

The most satisfying part of the experience has been to receive letters from readers who have enjoyed the book. I have tried to reply to each one, often continuing to discuss Einstein at length with them. A number of these readers sent me sources or new quotations, and some just wrote to share their enthusiasm for the book or for Einstein. But the majority, by far, have asked for the sources and contexts of quotations they have loved for many years. I was not always able to help them, and perhaps for good reason. I discovered that there appear to be many gremlins out there who are tagging Einstein's name on to lofty—and not so lofty—words they conjoin in the name of some cause or idea they're trying to promote. Some of these clever concoctions are listed in the "Attributed to Einstein" section at the back of the book. Readers of the first edition

Young Einstein scholar Brian Claeys in Iowa City in 1998,
meeting former Israeli foreign minister Abba Eban,
who once offered Israel's presidency to Einstein.
Congressman Jim Leach is at center. (Courtesy of the
Claeys family)

may notice that many of the quotations that were in
the original list in this section are no longer there
and can be found, documented, in the body of the
book.

One of my correspondents has been a young
farm boy from Iowa named Brian Claeys, and I
want to tell you about him because he delighted me

so much. Brian was thirteen years old when he contacted me in early 1998, at a time when he was putting together a school project on Einstein for National History Day, and he enlisted my help. I was impressed by his maturity, intelligence, and curiosity, and by the quality of the work he had already done. Before long, we had a lively correspondence going, and he was calling me by my first name (I had taken the liberty of calling him by his first name right away). He also wrote to Freeman Dyson and Stephen Hawking, among other physicists. He wrote to and met Abba Eban, the first Israeli ambassador to the United Nations and the United States and former foreign minister of Israel; he sent me a picture of them together with Brian's congressman, Jim Leach. He worked diligently and enthusiastically. In the end, his project took first place in his school, then in his region, then in his state, and next he planned to come to the Washington, D.C., area for the national finals. Because he was going to be relatively close by, I invited him to come north to Princeton with his family so he could see where Einstein lived and worked. He called to say he was coming with his mother. Soon after, on a warm spring morning, I went to the train station to welcome to Princeton Mrs. Claeys and a handsome American kid wearing a baseball cap. Over lunch at the faculty club, we spoke about Einstein and about growing soybeans and corn in the heart of the country. Next, I gave them my "Einstein's

Princeton" tour. I'm not sure if Princeton measured up to the farm back in Iowa, but I hope it planted some splendid memories in Brian's mind. A few months later I received a large, attractive basket in the mail, specially handmade for me by Brian himself. Incidentally, he had placed seventh in the national finals, but with me he definitely placed first.

The popularity of the original edition, boosted by translations into eighteen languages, has taken us by pleasant surprise. It shows that Einstein, the "lone wolf" whom *Time* magazine named "Man of the Century" at the turn of the millennium, is still celebrated as a cultural icon worldwide, and people everywhere are eager to find readable, factual information about him.

While the first edition was in the bookstores, I continued to research, read, and consult the materials in the Einstein Archive and kept my eyes and ears open for "new" quotations. The result is this enlarged edition, inching itself closer to being a kind of Einstein concordance. I have also made corrections and additions to some of the old quotations and sources, retranslated awkward passages, and expanded some annotations. I consulted the original German editions of some books, such as the one containing the Einstein-Born letters, and translated from them directly, though I give page numbers of the English editions, which are more readily available. I have added a section on music, an appendix, and new information that has come to light over the past three years. All new material is preceded by an asterisk (*).

Most of the acknowledgments in the preface still apply to this new edition of the book. In addition, I would like to thank Anneli Mynttinen, editorial associate at the Einstein Papers Project in Boston, for supplying me with some new material and references; Sam Schweber for thoughtfully providing me with photos and a boxful of Einstein material to browse through; Gene Dannen, for taking a keen interest, giving me several sources of quotations, and helping with a photo search during a time when he was very busy; Sandra Gebekken of the Leo Baeck Institute in New York, for her congeniality when I visited and her efficient service when I requested some photos; and the many people who, along with their friendly and gracious letters, sent me new quotations and copies of Einstein correspondence that I had not seen before.

Finally, I once again thank Robert Schulmann, director of the Einstein Papers Project, for being the country's best and most reliable source on Einstein; my valued friend Janet Stern for agreeing to edit the expanded edition, as well—in record time—with her usual expertise, flair, and finesse; Trevor Lipscombe for sponsoring the book, his sharp humor always intact; Devra Kupor for overseeing the book's production with such care and interest; Gretchen Oberfranc for the outstanding composition of the book; Jan Lilly for the wonderful design; and my children, Denise and David, for continuing to make me proud of them.

Facing page: Einstein Family Tree

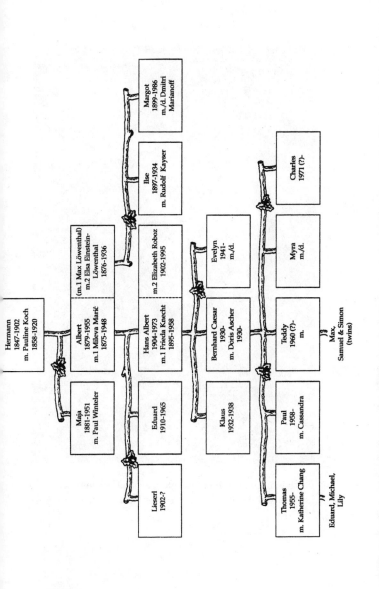

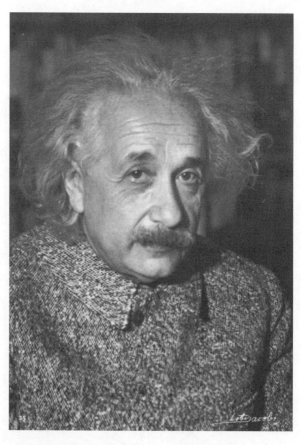

Einstein in middle age, Princeton, 1938. (Courtesy of
Lotte Jacobi Archives, University of New Hampshire)

Chronology

This chronology was assembled primarily from information contained in the chronologies of volumes 1 and 5 of the *Collected Papers of Albert Einstein*; from the chronology in *Subtle Is the Lord* by Abraham Pais; and from my notes and conversations with Helen Dukas in 1978–1980. It was supplemented by information gathered from additional readings.

1879   March 14, Albert Einstein is born in Ulm, Germany, in the home of his parents, Hermann (1847–1902) and Pauline Koch (1858–1920) Einstein.

1880   Family moves to Munich.

1881   November 18, Einstein's sister Maja is born.

1884   Receives from his father a compass, which makes a great impression on the young child.

1885   In the fall, enters the Petersschule, a Catholic primary school, where he is the only Jew in class. Receives Jewish religious instruction at home and becomes curious about religion; his religiosity ends by age twelve. Begins violin lessons.

1888   Enters Luitpold-*Gymnasium* in Munich.

1889–1895   Interest in physics, mathematics, and philosophy develops.

1894   Family moves to Italy, but Albert stays in Munich to finish school. He quits the *Gymnasium* at the end of the year and joins his family in Italy.

1895   Attempts to enter the Federal Polytechnical Institute (now the ETH—Eidgenössische Technische Hochschule) in Zurich in the fall, two years before the regular age of admission, but fails the entrance exam. Instead, attends the Aargau Cantonal School in Aarau (Aargau is the canton, Aarau the town on the right bank of the Aar River) while living in the home of one of his teachers, Jost Winteler, and his family.

1896   Relinquishes his German citizenship because his dislike of the German military mentality and remains stateless for the next five years. In the fall, is graduated from the Aargau school, entitling him to enter the Federal Polytechnical Institute, and he moves to Zurich at the end of October.

1899   Applies for Swiss citizenship at the age of twenty.

1900   Is graduated from the Polytechnical Institute, but his application to become an assistant at the Poly for the fall semester is turned down. In the summer, tells his disapproving mother that he plans to marry fellow student Mileva Marić. At end of year, sends his first scientific paper to the journal *Annalen der Physik*.

1901   Becomes a Swiss citizen. Seeks employment. His first scientific paper, "Conclusions Drawn from the Phenomena of Capillarity," is published in March in *Annalen der Physik*. In the summer, works as a substitute teacher at the technical school in Winterthur, and in the fall as a tutor in a private boarding school in Schaffhausen. Stays in touch with and visits Mileva regularly. Begins work on a doctoral dissertation on molecular forces in gases, which he submits to the University of Zurich in November. December, applies for a position at the Swiss Patent Office in Bern.

1902   Probably in January, daughter Lieserl is born out of wedlock to Mileva. Withdraws his doctoral dissertation from the University of Zurich. June, begins a provisional appointment as Technical Expert, Third Class, at the Patent Office in Bern. October, father dies in Milan.

1903   January 6, marries Mileva in Bern, where they take up residence. September, daughter Lieserl is registered, which may have indicated intention to put her up for adoption in case knowledge of the illegitimacy would be a threat to Einstein's federal appointment. No mention is made of Lieserl after she contracts scarlet fever in September while Mileva is on a visit to Budapest. (It appears that Lieserl never lived with her parents, Einstein never saw his daughter, and all trace of her has been lost.) At this time, Mileva is pregnant again.

1904   May 14, son Hans Albert ("Adu") is born in Bern (died 1973 in Falmouth, Massachusetts; buried in Woods Hole, Mass.). September, Einstein's provisional appointment at the Patent Office becomes permanent.

1905   Einstein's "year of miracles" with respect to his scientific publications. April 30, submits his doctoral dissertation, "A New Determination of Molecular Dimensions," for publication. In addition, publishes three of his most important scientific papers: "On a Heuristic Point of View Concerning the Production and Transformation of Light" (published June 9), which deals with the quantum hypothesis, showing that electromagnetic radiation interacts with matter as if the radiation has a granular structure (the so-called photoelectric effect); "On the Movement of Small Particles Suspended in Stationary Liquids Required by the Molecular-Kinetic Theory of Heat" (published July 18), his first paper on Brownian motion, leading to experiments validating the kinetic-molecular theory of heat; and "On the Electrodynamics of Moving Bodies" (published September 26), his first paper on the special theory of relativity and a landmark in the development of modern physics. A second, shorter paper on the special theory, published November 21, contains the relation $E = mc^2$ in its original form (see the quotation under $E = mc^2$ in the section "On Science and Scientists, Mathematics, and Technology").

1906   January 15, formally receives doctorate from the University of Zurich. March 10, promoted to Technical Expert, Second Class, at the Patent Office.

1907   While still at the Patent Office, seeks other employment, including at the cantonal school in Zurich and at the University of Bern.

1908   February, becomes a *Privatdozent* (lecturer) at the University of Bern. Sister Maja receives her doctorate in romance languages from the University of Bern.

1909   May 7, is appointed Extraordinary Professor of Theoretical Physics at the University of Zurich, effective October 15. Resigns from his positions at the Swiss Patent Office and the University of Bern. Receives his first honorary doctorate, from the University of Geneva.

1910   March, sister Maja marries Paul Winteler, son of Einstein's teacher in Aargau. July 28, second son,

Eduard ("Tete"), is born (died 1965 in a psychiatric hospital in Burghölzli, Switzerland; he had had a history of schizophrenia since he was in his twenties). October, completes a paper on critical opalescence and the blue color of the sky, his last major work in classical statistical physics.

1911   Accepts an appointment as director of the Institute of Theoretical Physics at the German University of Prague, effective April 1, and resigns his position at the University of Zurich. Moves his family to Prague. October 29, attends the first Solvay Congress in Brussels.

1912   Becomes acquainted with his divorced cousin Elsa Löwenthal and begins a romantic correspondence with her as his own marriage disintegrates. Accepts appointment as Professor of Theoretical Physics at the ETH in Zurich, beginning in October, and resigns his position in Prague.

1913   September, sons Hans Albert and Eduard are baptized as Orthodox Christians near Novi Sad, Yugoslavia, their mother's hometown. November, is elected to the Prussian Academy of Sciences and is offered a position in Berlin, home of Elsa Löwenthal. The offer includes a research professorship at the University of Berlin, without teaching obligations, and the directorate of the soon-to-be-established Kaiser Wilhelm Institute of Physics. Resigns from the ETH.

1914   April, arrives in Berlin to assume his new position. Mileva and the children join him but soon return to Zurich because of Mileva's unhappiness in Berlin. August, World War I begins.

1915   Co-signs a "Manifesto to Europeans" upholding European culture, probably his first public

political statement. November, completes his
work on the logical structure of general relativity.

1916   Publishes "The Origins of the General Theory of
Relativity" (later to become his first book) in *Annalen der Physik*. May, becomes president of the
German Physical Society. Publishes three papers
on quantum theory.

1917   February, writes his first paper on cosmology. Becomes ill and is weakened by a liver ailment and
an ulcer. Elsa takes care of him. October 1, begins
directorship of the Kaiser Wilhelm Institute of
Physics. After World War I, holds dual Swiss and
German citizenship.

1919   February 14, is divorced from Mileva. Divorce de-
cree stipulates that any future Nobel Prize monies
go to her and the children for living expenses
and to ensure their permanent financial security.
May 29, during a solar eclipse, Sir Arthur Edding-
ton experimentally measures the bending of light
and confirms Einstein's predictions; Einstein's
fame as a public figure begins. June 2, marries
Elsa, who has two unmarried daughters, Ilse
(22 years old) and Margot (20 years old), living
at home. Late in the year, becomes interested
in Zionism through his friendship with Kurt
Blumenfeld.

1920   March, mother dies in Berlin. Expressions of anti-
Semitism and anti–relativity theory become no-
ticeable among Germans, yet Einstein remains
loyal to Germany. Becomes increasingly involved
in nonscientific interests.

1921   April and May, makes first trip to the United
States. Receives honorary degree and delivers
four lectures on relativity theory at Princeton

University as part of the Stafford Little Lectures, which Princeton University Press in the United States and Methuen and Company in Great Britain later publish as *The Meaning of Relativity*. Accompanies Chaim Weizmann on U.S. fund-raising tour on behalf of Hebrew University of Jerusalem.

1922   Completes his first paper on a unified field theory. October through December, takes trip to Japan, with other stops en route to the Far East. November, while in Shanghai, learns that he has won the 1921 Nobel Prize in physics (see "Answers to Common Questions," at the back of the book).

1923   Visits Palestine and Spain.

1924   Stepdaughter Ilse marries Rudolf Kayser, a journalist and subsequent Einstein biographer. Ilse had, for a time, considered marrying Einstein, who appears to have been in love with her, before he married her mother.

1925   Travels to South America. In solidarity with Gandhi, signs a manifesto against compulsory military service. Becomes an ardent pacifist. Receives Copley Medal. Until 1928, serves on Board of Governors of Hebrew University.

1926   Royal Astronomical Society of England awards him its gold medal.

1927   Son Hans Albert marries Frieda Knecht.

1928   Falls ill again, this time with a heart problem. Is confined to bed for several months and remains weak for a year. April, Helen Dukas is hired as his secretary and remains with him as secretary and housekeeper for the rest of his life.

1929   Begins lifelong friendship with Queen Elizabeth of Belgium. June, receives Planck Medal.

1930   First grandchild, Bernhard, is born to Hans Albert and Frieda. Stepdaughter Margot marries Dmitri Marianoff (marriage later ends in divorce). Signs manifesto for world disarmament. December, visits New York and Cuba and stays (until March 1931) at the California Institute of Technology (Caltech), in Pasadena.

1931   Visits Oxford in May to deliver the Rhodes Lectures and receives honorary degree, then spends several months at his summer cottage in Caputh, southwest of Berlin. December, en route to Pasadena again.

1932   January–March, visits Caltech again. Returns to Berlin. Later, agrees to accept an appointment as professor at the Institute for Advanced Study in Princeton, to begin when its campus is completed late in the decade. December, makes another visit to the United States.

1933   January, Nazis come to power. Resigns membership in the Prussian Academy of Sciences, gives up German citizenship (remains a Swiss citizen), and does not return to Germany. Instead, from the U.S., goes to Belgium with Elsa and sets up temporary residence at Coq-sur-Mer. Ilse, Margot, Helen Dukas, and Walther Mayer, an assistant, join them, and security guards are assigned to protect them. Takes trips to Oxford, where he delivers the Herbert Spencer Lecture in June, and Switzerland, where he makes what will be his final visit to son Eduard. Rudolf Kayser, Ilse's husband, manages to have Einstein's papers in Berlin sent to France and eventually brought to the United States. September, leaves Europe, together with Elsa, Helen Dukas, and Walther Mayer, and arrives in New York on October 17 on the *Westmoreland*; Ilse and Margot and their spouses remain in Europe. Publishes, with Sigmund Freud, *Why War?* Begins professorship at the Institute for Advanced Study, temporarily located in the old Fine Hall (now Jones Hall) on the Princeton University campus.

1934   July 10, Ilse dies in Paris at age 37 after a long and painful illness. Margot and Dmitri come to Princeton.

1935   Fall, moves to 112 Mercer Street, Princeton, where Einstein, Elsa, Margot, Maja, and Helen Dukas will live out their lives. Receives Franklin Medal.

1936   Hans Albert receives doctorate in technical sciences from the ETH in Zurich (in 1947 he becomes a professor of hydraulic engineering at the University of California at Berkeley). December 20, Elsa dies after a long battle with heart and kidney disease.

1939   Sister, Maja Winteler-Einstein, comes to live at Mercer Street. August 2, signs famous letter to President Roosevelt on the military implications of atomic energy (see Appendix). World War II begins in Europe.

1940   Receives U.S. citizenship. Maintains dual U.S. and Swiss citizenship until his death. Citizenship had been proposed earlier by an act of Congress, but Einstein preferred waiting to be naturalized the customary way.

1941   December, United States enters World War II.

1943   Becomes consultant to U.S. Navy Bureau of Ordnance, Section on Explosives and Ammunition.

1944   A newly handwritten copy of the original 1905 paper on the special theory of relativity is auctioned off for $6 million as a contribution to the war effort.

1945   World War II ends. Retires officially from the faculty of the Institute for Advanced Study, receives a pension, but continues to keep an office there until his death.

1946   Maja suffers a stroke and is confined to bed. Einstein becomes chairman of the Emergency Committee of Atomic Scientists. Urges United Nations to form a world government, declaring that it is the only way to maintain world peace.

1948   August 4, Mileva dies in Zurich. December, Einstein's doctors tell him that he has a large aneurysm (abnormal dilatation) of the abdominal aorta.

1950   March 18, signs his last will, naming his friend Otto Nathan as executor and Otto Nathan and Helen Dukas as trustees of his estate. His literary estate (the archive) is to be transferred to the Hebrew University of Jerusalem after the death of Nathan and Dukas. (Arrangements are later made for an earlier transfer.)

1951   June, Maja dies in Princeton.

1952   Is offered the presidency of Israel, which he declines.

1954   Develops hemolytic anemia.

1955   April 11, writes last signed letter, to Bertrand Russell, agreeing to sign a joint manifesto urging all nations to renounce nuclear weapons. April 13, aneurysm ruptures. April 15, enters Princeton Hospital. April 18, Albert Einstein dies at 1:15 A.M. of a ruptured arteriosclerotic aneurysm of the abdominal aorta, caused by hardening of the arteries. He had opposed surgery, but the autopsy showed it would not have helped.

1965   Eduard Einstein dies.

1973   Hans Albert Einstein dies.

1982   January, Helen Dukas dies.

1986   July 8, Margot Einstein dies.

1987   Volume 1 of *The Collected Papers of Albert Einstein* is published.

Quotable Einstein

On Einstein Himself

Einstein in fuzzy slippers. (Courtesy of Gillett Griffin, Princeton)

A happy man is too satisfied with the present to think too much about the future.

Written at age seventeen (September 18, 1896) for a school French essay entitled "My Future Plans." *CPAE*, Vol. 1, Doc. 22

* Strenuous intellectual work and the study of God's Nature are the angels that will lead me through all the troubles of this life with consolation, strength, and uncompromising rigor.

To Pauline Winteler, mother of Einstein's girlfriend Marie, ca. May 1897. *CPAE*, Vol. 1, Doc. 34

I decided the following about our future: I will look for a position *immediately*, no matter how humble it is. My scientific goals and my personal vanity will not prevent me from accepting even the most subordinate position.

To future wife Mileva Marić, July 7, 1901, while having difficulty finding his first job. *CPAE*, Vol. 1, Doc. 114

I have come to know the mutability of all human relationships and have learned to insulate myself against both heat and cold so that a temperature balance is fairly well assured.

> To Heinrich Zangger, March 10, 1917. *CPAE*,
> Vol. 8, Doc. 309

* I am by heritage a Jew, by citizenship a Swiss, and by makeup a human being, and *only* a human being, without any special attachment to any state or national entity whatsoever.

> To Alfred Kneser, June 7, 1918. *CPAE*, Vol. 8,
> Doc. 560

* I was originally supposed to become an engineer, but the thought of having to expend my creative energy on things that make practical everyday life even more refined, with a loathsome capital gain as the goal, was unbearable to me.

> To Heinrich Zangger, ca. August 11, 1918. *CPAE*,
> Vol. 8, Doc. 597

Here is yet another application of the principle of relativity . . . : today I am described in Germany as a "German savant" and in England as a "Swiss Jew." Should it ever be my fate to be represented as a bête noire, I should, on the contrary, become a "Swiss Jew" for the Germans and a "German savant" for the English.

To *The Times* (London), 1919. Quoted in Hoffmann,
Albert Einstein: Creator and Rebel, 139; also quoted
in Frank, *Einstein: His Life and Times*, 144

* I have not yet eaten enough of the Tree of Knowledge, though in my profession I am obliged to feed on it regularly.

To Max Born, November 9, 1919. In Born, *Born-Einstein Letters*, 16

With fame I become more and more stupid, which of course is a very common phenomenon.

To Heinrich Zangger, December 1919. Einstein
Archive 39-726; also quoted in Dukas and
Hoffmann, *Albert Einstein, the Human Side*, 8

* My father's ashes lie in Milan. I buried my mother here [Berlin] only a few days ago. My children are in Switzerland. . . . I myself have journeyed everywhere continuously—a stranger everywhere. . . . A person like me is at home anywhere with those near and dear to him.

To Max Born, March 3, 1920. In Born, *Born-Einstein Letters*, 26

Let me tell you what I look like: pale face, long hair, and a tiny start of a paunch. In addition, an awkward gait, and a cigar in the mouth . . . and a pen in pocket or hand. But crooked legs and warts he does not have, and so is quite handsome—also there's no hair on his hands, as is so often the case with ugly men. So it really is a pity that you didn't see me.

Postcard to eight-year-old cousin Elisabeth Ney, September 1920. Einstein Archive 36-525; also quoted in Dukas and Hoffmann, *Albert Einstein, the Human Side*, 44

Just as with the man in the fairy tale who turned whatever he touched into gold, with me everything is turned into newspaper clamor.

To Max Born, September 9, 1920. Einstein Archive 8-151

Personally, I experience the greatest degree of pleasure in having contact with works of art. They furnish me with happy feelings of an intensity that I cannot derive from other sources.

1920. In Moszkowski, *Conversations with Einstein*, 184

It strikes me as unfair, and even in bad taste, to select a few individuals for boundless admiration, attributing superhuman powers of mind and character to them. This has been my fate, and the contrast between the popular assessment of my powers and achievements and the reality is simply grotesque.

From an interview, *Nieuwe Rotterdamsche Courant*,
1921; reprinted in *Ideas and Opinions*, 3–7

If my theory of relativity is proven successful, Germany will claim me as a German and France will
declare that I am a citizen of the world. Should my
theory prove untrue, France will say that I am a
German and Germany will declare that I am a Jew.

From an address to the French Philosophical
Society at the Sorbonne, April 6, 1922. See also
French press clipping, April 7, 1922, Einstein
Archive 36-378; and *Berliner Tageblatt*, April 8,
1922, Einstein Archive 79-535

* When a blind beetle crawls over the surface of a
curved branch, it doesn't notice that the track it has
covered is indeed curved. I was lucky enough to
notice what the beetle didn't notice.

In answer to his son Eduard's question about why
he is so famous, 1922. Quoted in Max Flückiger,
Albert Einstein in Bern (Bern: Haupt, 1961); also
quoted in Grüning, *Ein Haus für Albert Einstein*, 498

* Now I am sitting peacefully in Holland after being
told that certain people in Germany have it in for
me as a "Jewish saint." In Stuttgart there was even
a poster in which I appeared in first place among
the richest Jews.

To sons Hans Albert and Eduard, November 24,
1923

* [I] must seek in the stars that which was denied [to me] on earth.

> To his secretary Bette Neumann, ca. 1923–1924,
> with whom he had fallen in love, upon ending his
> relationship with her. See Pais, *Subtle Is the Lord*,
> 320, and Fölsing, *Albert Einstein*, 548

* Of all the communities available to us, there is not one I would want to devote myself to except for the society of the true seekers, which has very few living members at any one time.

> To Max and Hedwig Born, April 29, 1924. In Born,
> *Born-Einstein Letters*, 82

* Imagination is more important than knowledge. For knowledge is limited, whereas imagination embraces the entire world, stimulating progress, giving birth to evolution.

> Originally in "What Life Means to Einstein,"
> *Saturday Evening Post*, October 26, 1929; reprinted
> in "On Science," in *Cosmic Religion*, 97.

To punish me for my contempt of authority, Fate has made me an authority myself.

> Aphorism for a friend, September 18, 1930.
> Einstein Archive 36–598; also quoted in Hoffmann,
> *Albert Einstein: Creator and Rebel*, 24

I am an artist's model.

> To a fellow train passenger, October 31, 1930, who
> asked him his occupation, reflecting Einstein's
> feeling that he was constantly posing for
> sculptures and paintings. Einstein Archive 21-006;
> also quoted in ibid., 4

I have never looked upon ease and happiness as ends in themselves—such an ethical basis I call the ideal of a pigsty. . . . The ideals which have guided my way, and time after time have given me the energy to face life, have been Kindness, Beauty, and Truth.

> From "What I Believe," *Forum and Century* 84
> (1930), 193–194; reprinted in *Ideas and Opinions*,
> 8–11

I am truly a "lone traveler" and have never belonged to my country, my home, my friends, or even my immediate family, with my whole heart. In the face of all this, I have never lost a sense of distance and the need for solitude.

> Ibid. Sometimes translated as "I am a lone wolf"
> and "I am a horse for a single harness."

A hundred times a day I remind myself that my inner and outer lives are based on the labors of other people, living and dead, and that I must exert myself in order to give in the same measure as I have received and am still receiving.

Ibid.

It is an irony of fate that I myself have been the recipient of excessive admiration and reverence from my fellow-beings, through no fault or merit of my own.

Ibid.

Professor Einstein begs you to treat your publications for the time being as if he were already dead.

Written on Einstein's behalf by his secretary, Helen Dukas, March 1931, after he was besieged by one manuscript too many. Einstein Archive 46-487

Although I am a typical loner in my daily life, my awareness of belonging to the invisible community of those who strive for truth, beauty, and justice has prevented me from feelings of isolation.

From "My Credo," for the German League for Human Rights, 1932. Quoted in Leach, *Living Philosophies*, 3

Although I try to be universal in thought, I am European by instinct and inclination.

Daily Express (London), September 11, 1933. Quoted in Holton, *Advancement of Science*, 126

People flatter me so long as I don't get in their way. [At other times] they immediately turn to abuse and calumny in defense of their interests.

> To a pacifist friend. Published in *Mein Weltbild* (1934), 54; reprinted in *Ideas and Opinions*, 110

* To be called to account publicly for what *others* have said in your name, when you cannot defend yourself, is a sad situation indeed.

> From "Interviewers," in ibid., 40 and 15, respectively

* My life is a simple thing that would interest no one. It is a known fact that I was born, and that is all that is necessary.

> To Princeton High School reporter Henry Russo. In *The Tower*, April 13, 1935

* As a boy of twelve years making my acquaintance with elementary mathematics, I was thrilled in seeing that it was possible to find out truth by reasoning alone, without the help of any outside experience. . . . I became more and more convinced that even nature could be understood as a relatively simple mathematical structure.

> Ibid.

I have acclimated extremely well here, live like a bear in its cave, and feel more at home than ever before in my eventful life. This bearlike quality has increased even more because of the death of my mate, who was more attached to other people than I am.

> To Max Born, ca. early 1937, after the death of
> Einstein's wife, Elsa. In Born, *Born-Einstein Letters*,
> 128

I wouldn't want to live if I did not have my work. . . . In any case, it's good that I'm already old and personally don't have to count on a prolonged future.

> To close friend Michele Besso, October 10, 1938,
> reflecting on Hitler's rise to power. Einstein
> Archive 7-376

* I firmly believe that love [of a subject or hobby] is a better teacher than a sense of duty—at least for me.

> Draft of a letter to Philipp Frank, 1940

Why is it that nobody understands me, yet everybody likes me?

> From an interview, *New York Times*, March 12, 1944

* I do not like to state an opinion on a matter unless I know the precise facts.

From an interview with Richard J. Lewis, *New York Times*, August 12, 1945, 29:3, on declining to comment on Germany's progress on the atom bomb

I never worry about the future. It comes soon enough.

Aphorism, 1945–46. Einstein Archive 36-570

I have to apologize to you that I am still among the living. There *will* be a remedy for this, however.

To a child, Tyffany Williams, in South Africa, August 25, 1946, after she expressed surprise in a letter that Einstein was still alive. Einstein Archive 42-612

* What is essential in the life of a man of my kind is *what* he thinks and *how* he thinks, and not what he does or suffers.

Written in 1946 for "Autobiographical Notes," in Schilpp, *Albert Einstein: Philosopher-Scientist*, 33

There have already been published by the bucketsful such brazen lies and utter fictions about me that I would long since have gone to my grave if I had allowed myself to pay attention to them.

To the writer Max Brod, February 22, 1949. Einstein Archive 34-066

My scientific work is motivated by an irresistible longing to understand the secrets of nature and by no other feelings. My love for justice and the striving to contribute toward the improvement of human conditions are quite independent from my scientific interests.

To F. Lentz, August 20, 1949, in answer to a letter asking Einstein about his scientific motivation. Einstein Archive 58-418

* I'm doing just fine, considering that I have triumphantly survived Nazism and two wives.

To Jakob Ehrat, May 12, 1952

It is a strange thing to be so widely known, yet to be so lonely. But it is a fact that this kind of popularity . . . is forcing its victim into a defensive position that leads to isolation.

To E. Marangoni, October 1, 1952. Einstein Archive 60-406

I have no special talents. I am only passionately curious.

To Carl Seelig, his biographer, March 11, 1952. Einstein Archive 39-013. A similar sentiment was expressed in a letter to Hans Muehsam, March 4, 1953, Einstein Archive 38-424

All my life I have dealt with objective matters; hence I lack both the natural aptitude and the experience to deal properly with people and to carry out official functions.

Statement to Abba Eban, Israeli ambassador to the United States, November 18, 1952, turning down the presidency of Israel after Chaim Weizmann's death. Einstein Archive 28-943

In the past it never occurred to me that every casual remark of mine would be snatched up and recorded. Otherwise I would have crept further into my shell.

To Carl Seelig, October 25, 1953. Einstein Archive 39-053

All manner of fable is being attached to my personality, and there is no end to the number of ingeniously devised tales. All the more do I appreciate and respect what is truly sincere.

To Queen Elizabeth of Belgium, March 28, 1954. Einstein Archive 32-410

I'm not the kind of snob or exhibitionist that you take me to be and furthermore have nothing of value to say of immediate concern, as you seem to assume.

In reply to a letter, May 1954, asking Einstein to send a message to a new museum in Chile, to be

put on display for others to admire. Einstein
Archive 60-624

* It is true that my parents were worried because I
began to speak fairly late, so that they even consulted a doctor. I can't say how old I was—but
surely not less than three.

> To Sybille Blinoff, May 21, 1954. Einstein Archive
> 59-261. In her biography of him, Einstein's sister,
> Maja, puts his age at two and a half. *CPAE*, Vol. 1,
> lvii

It is quite curious, even abnormal, that with your
superficial knowledge about the subject you are so
confident in your judgment. I regret that I cannot
spare the time to occupy myself with dilettantes.

> To dentist G. Lebau, who claimed he had a better
> theory of relativity, July 10, 1954. The dentist
> returned Einstein's letter with a note written at the
> bottom: "I am thirty years old; it takes time to
> learn humility." Einstein Archive 60-226

If I would be a young man again and had to decide
how to make my living, I would not try to become
a scientist or scholar or teacher. I would rather
choose to be a plumber or a peddler, in the hope of
finding that modest degree of independence still
available under present circumstances.

> To the editor, *The Reporter* magazine, October 13,
> 1954. Also quoted in Nathan and Norden, *Einstein*

on Peace, 613. To which a plumber, Stanley Murray,
replied to Einstein on November 11: "Since my
ambition has always been to be a scholar and
yours seems to be a plumber, I suggest that as a
team we would be tremendously successful. We
can then be possessed of both knowledge and
independence." Rosenkranz, *Albert through the
Looking-Glass*, 74–75. At other times Einstein
claimed that he would choose to be a musician or
a lighthouse keeper.

Only in mathematics and physics was I, through
self-study, far beyond the school curriculum, and
also with regard to philosophy as it was taught in
the school curriculum.

From a 1955 letter; quoted in Hoffmann, *Albert Einstein: Creator and Rebel*, 20

To me it is enough to wonder at the secrets.

Quoted in A&E Television's Einstein biography,
VPI International, 1991. This may be a paraphrase,
as I haven't been able to find it anywhere as stated.

Arrows of hate have been aimed at me too, but they
have never hit me, because somehow they belonged to another world with which I have no connection whatsoever.

Quoted in *Out of My Later Years*, 13

My intuition was not strong enough in the field of mathematics to differentiate clearly the fundamentally important . . . from the rest of the more or less desirable erudition. Also, my interest in the study of nature was no doubt stronger. . . . In this field I soon learned to sniff out that which might lead to fundamentals and to turn aside . . . from the multitude of things that clutter up the mind and divert from the essentials.

"Autobiographical Notes," in Schilpp, *Albert Einstein: Philosopher-Scientist*, 15

The only way to escape the corruptible effect of praise is to go on working.

Quoted by Lincoln Barnett, "On His Centennial, the Spirit of Einstein Abides in Princeton," *Smithsonian*, February 1979, 74

God gave me the stubbornness of a mule and a fairly keen scent.

Quoted in Whitrow, *Einstein: The Man and His Achievement*, 91

* The ordinary adult never gives a thought to space-time problems. . . . I, on the contrary, developed so slowly that I did not begin to wonder about space and time until I was an adult. I then delved more

deeply into the problem than any other adult or child would have done.

> To Nobel laureate James Franck, on his belief that it is usually children, not adults, who reflect on space-time problems. Quoted in Seelig, *Helle Zeit, dunkle Zeit*, 72

* I very rarely think in words at all. A thought comes, and I may try to express it in words *afterwards.*

> To psychologist Max Wertheimer. In Wertheimer, *Productive Thinking* (New York: Harper, 1959), 213–228

* The development of that mental world (*Gedanken-welt*) is a continual flight from "wonder." I experienced such a wonder when my father showed me a compass at the age of four or five.

> "Autobiographical Notes," in Schilpp, *Albert Einstein: Philosopher-Scientist*, 8–9

When I was young, all I wanted and expected from life was to sit quietly in some corner doing my work without the public paying attention to me. And now see what has become of me.

> Quoted in Hoffmann, *Albert Einstein: Creator and Rebel*, 4

When I examine myself and my methods of thought, I come close to the conclusion that the gift of imagination has meant more to me than my talent for absorbing absolute knowledge.

Recalled by a friend on the one hundredth anniversary of Einstein's birth, celebrated February 18, 1979. Quoted in Ryan, *Einstein and the Humanities*, 125

I have never obtained any ethical values from my scientific work.

Quoted in Michelmore, *Einstein: Profile of the Man*, 251

That little word "we" I mistrust and here's why:
No man of another can say he is I.
Behind all agreement lies something amiss
All seeming accord cloaks a lurking abyss.

Verse quoted in Dukas and Hoffmann, *Albert Einstein, the Human Side*, 100

I have finished my task here.

Said as he was dying. Einstein Archive 39-095. Taken from biographer Carl Seelig's account, who perhaps heard it from Helen Dukas.

Sons Eduard (*left*) and Hans Albert in Switzerland, circa 1918. (Courtesy of Leo Baeck Institute, New York)

According to Einstein, his marriage to Mileva, who
came from a Serbo-Greek peasant background,
lasted for seventeen years, but he never really knew
her. He recalled that he had married her primarily
"from a sense of duty," possibly because she had
given birth to their illegitimate child. "I had, with
an inner resistance, embarked on something that
simply exceeded my strength." At the time of their
marriage, he did not know that mental illness was
a hereditary disease on Mileva's mother's side of
the family, and both mother and daughter were
often depressed and suspicious. Mileva was also
physically disfigured—one leg was shorter than the
other—due to a congenital hip displacement, and
this deformity added to her emotional problems.
Unable to accept her eventual divorce and Einstein's
often insensitive treatment of her, she became
bitter, sometimes causing difficulties in Einstein's
relationship with his two sons. The many letters he
wrote to them, especially to Hans Albert, show that
he tried to remain close to them during their
childhood and was a warm and caring father. He
also eventually conceded that Mileva was a good
mother. (See *CPAE*, Vol. 8, for these letters as well as
letters to Mileva in which the couple tries to deal
with its financial and parenting difficulties after
the separation.) Still, these tragic circumstances,
according to Einstein, left their mark on him into
his old age and may have amplified his deep

involvement in impersonal things. See letters to his biographer Carl Seelig, March 26 and May 5, 1952; Einstein Archive 39-016 and 39-020

Mama threw herself on the bed, buried her head in the pillow, and wept like a child. After regaining her composure, she immediately shifted to a desperate attack: "You're ruining your future and destroying your opportunities." "No decent family would want her." "If she becomes pregnant, you'll be in a real mess." With this outburst, which was preceded by many others, I finally lost my patience.

To Mileva, July 29, 1900, after telling his mother that he and Mileva planned to marry; they did not marry until January 6, 1903. *The Love Letters*, 19; *CPAE*, Vol. 8, Doc. 68

I long terribly for a letter from my beloved witch. I find it hard to believe that we will be separated for so much longer—only now do I see how much in love with you I am! Pamper yourself, so you will become a radiant little sweetheart and as wild as a street urchin!

To Mileva, August 1, 1900. *The Love Letters*, 21; *CPAE*, Vol. 1, Doc. 69

* When you're not with me, I feel as though I'm not complete. When I'm sitting, I want to go away;

when I go away, I'd rather be home; when I'm talking with people, I'd rather be studying; when I study, I can't sit still and concentrate; and when I go to sleep, I'm not satisfied with the way the day has passed.

To Mileva, August 6, 1900. *The Love Letters*, 23–24; *CPAE*, Vol. 1, Doc. 70

How was I ever able to live alone, my little everything? Without you I have no self-confidence, no passion for work, and no enjoyment of life—in short, without you, my life is a void.

To Mileva, ca. August 14, 1900. *The Love Letters*, 26; *CPAE*, Vol. 1, Doc. 72

My parents are very concerned about my love for you. . . . They cry for me almost as if I had already died. Again and again they complain that I brought misfortune on myself by my devotion to you.

To Mileva, August–September 1900. *The Love Letters*, 29; *CPAE*, Vol. 1, Doc. 74

Without the thought of you, I would no longer want to live among this sorry herd of humans. But having you makes me proud, and the thought of you makes me happy. I will be doubly happy when I can press you to my heart once again and see those loving eyes shine for me alone, and when

I can kiss that sweet mouth that trembles for me only.

> Ibid.

I am also looking forward to working on our new studies. You must continue with your research— how proud I will be to have a little Ph.D. for a sweetheart while I remain a totally ordinary person!

> To Mileva, September 13, 1900. *The Love Letters*, 32;
> *CPAE*, Vol. 1, Doc. 75

Shall I look around for possible jobs for you [in Zurich]? I think I'll try to find some tutoring positions that I can later turn over to you. Or do you have something else in mind? . . . No matter what happens, we'll have the most wonderful life imaginable.

> To Mileva, September 19, 1900. *The Love Letters*, 33;
> *CPAE*, Vol. 1, Doc. 76

I am so lucky to have found you—a creature who is my equal, and who is as strong and independent as I am.

> To Mileva, October 3, 1900. *The Love Letters*, 36;
> *CPAE*, Vol. 1, Doc. 79

* How happy and proud I will be when the two of us together have brought our work on relative motion to a triumphant end!

> To Mileva, March 27, 1901. *The Love Letters*, 29;
> *CPAE*, Vol. 1, Doc. 94. See the discussion "Mileva
> Marić as Collaborator" in "Answers to Common
> Questions" at the back of the book.

You'll see for yourself how pleasant and cheerful I've become and how all of my scowling is a thing of the past. And I love you so much again! It was only because of nervousness that I was so mean to you . . . and I'm longing so much to see you again.

> To Mileva, April 30, 1901. *The Love Letters*, 46;
> *CPAE*, Vol. 1, Doc. 102

If only I could give you some of my happiness so you would never be sad and depressed again.

> To Mileva, May 9, 1901. *The Love Letters*, 51; *CPAE*,
> Vol. 1, Doc. 106

My wife is coming to Berlin with very mixed feelings because she is afraid of the relatives, probably mostly of you. . . . But you and I can be very happy with each other without her having to be hurt. You can't take away from her something she doesn't have [i.e., his love].

> To newfound love, cousin Elsa Löwenthal, August
> 1913. *CPAE*, Vol. 5, Doc. 465

The situation in my house is ghostlier than ever: icy silence.

> To Elsa, October 16, 1913. *CPAE*, Vol. 5, Doc. 478

Do you think it's so easy to get a divorce when one has no proof of the other party's guilt? . . . I am treating my wife like an employee whom I can't fire. I have my own bedroom and avoid being with her. . . . I don't know why you're so terribly upset by all of this. I'm absolutely my own master . . . as well as my own wife.

> To Elsa, before December 2, 1913. *CPAE*, Vol. 5, Doc. 488

[My wife, Mileva] is an unfriendly, humorless creature who gets nothing out of life and who, by her mere presence, extinguishes other people's joy of living.

> To Elsa, after December 2, 1913. *CPAE*, Vol. 5, Doc. 489

My wife whines incessantly to me about Berlin and her fear of the relatives. . . . My mother is good-natured, but she is a really fiendish mother-in-law. When she stays with us, the air is full of dynamite. . . . But both are to be blamed for their miserable relationship. . . . No wonder that my scientific life thrives under these circumstances: it lifts me impersonally from the vale of tears into a more peaceful atmosphere.

To Elsa, after December 21, 1913. *CPAE*, Vol. 5, Doc. 497

* (A) You will see to it that (1) my clothes and laundry are kept in good order; (2) I will be served three meals regularly *in my room*; (3) my bedroom and study are kept tidy, and especially that my desk is left for *my use only*. (B) You will relinquish all personal relations with me insofar as they are not completely necessary for social reasons. Particularly, you will forgo my (1) staying at home with you; (2) going out or traveling with you. (C) You will obey the following points in your relations with me: (1) you will not expect any tenderness from me, nor will you offer any suggestions to me; (2) you will stop talking to me about something if I request it; (3) you will leave my bedroom or study without any back talk if I request it. (D) You will undertake not to belittle me in front of our children, either through words or behavior.

Memorandum to Mileva, ca. July 18, 1914, listing the conditions under which he would agree to continue to live with her in Berlin. At first she accepted the conditions, but then left Berlin with the children at the end of July. *CPAE*, Vol. 8, Doc. 22

I don't want to lose the children, and I don't want them to lose me. . . . After everything that has happened, a friendly relationship with you is out of the question. We shall have a considerate and business-like relationship. All personal things must be kept to a minimum. . . . I don't expect I'll ask you for a divorce but only want you to stay in Switzerland with the children . . . and send me news of my precious boys every two weeks. . . . In return, I assure you of proper comportment on my part, such as I would exercise toward any unrelated woman.

> To Mileva, ca. July 18, 1914, on his offer to
> continue their marriage after his move to Berlin, to
> which in the end she did not agree. *CPAE*, Vol. 8,
> Doc. 23

* I came to realize that living with the children is no blessing if the wife stands in the way.

> To Elsa, July 26, 1914. *CPAE*, Vol. 8, Doc. 26

* I may see my children only on neutral ground, not in our [future] home. This is justified because it is not right to have the children see their father with a woman other than their own mother.

> To Elsa, after July 26, 1914. *CPAE*, Vol. 8, Doc. 27

* How much I look forward to the quiet evenings we'll be able to spend chatting alone, and to all the peaceful shared experiences still ahead of us! Now, after all my deliberations and work I'll find a precious little wife at home who receives me with cheer and contentment. . . . It wasn't her [Mileva's] ugliness, but her obstinacy, inflexibility, stubbornness, and insensitivity that prevented harmony between us.

To Elsa, July 30, 1914. *CPAE*, Vol. 8, Doc. 30

* There are reasons why I could not endure being with this woman any longer, despite the tender love that ties me to the children.

To Heinrich Zangger, November 26, 1915. *CPAE*, Vol. 8, Doc. 152

* You have no idea of the natural craftiness of such a woman. I would have been physically and mentally broken if I had not finally found the strength to keep her at arm's length and out of sight and earshot.

To Michele Besso, July 14, 1916. *CPAE*, Vol. 8, Doc. 233

* She leads a worry-free life, has her two precious boys with her, lives in a fabulous neighborhood,

does what she likes with her time, and innocently stands by as the guiltless party.

> To Michele Besso, July 21, 1916. *CPAE*, Vol. 8, Doc. 238

* The only thing she is missing is someone to dominate her. . . . What *man* would tolerate something so palpably smelly being stuck up his nose all his life, for no purpose at all, with the secondary obligation of also putting on a friendly face?

> Ibid.

* From now on I will no longer bother her about a divorce. The accompanying battle with my relatives has taken place. I have learned to withstand the tears.

> To Michele Besso, September 6, 1916. Einstein's relatives did not approve of his leaving his marriage in limbo, feeling it would compromise young Ilse's (Elsa's elder daughter's) eligibility for marriage. The divorce finally did take place in February 1919. Einstein, as the guilty party, was ordered not to marry for the next two years; but, despite the ban, he married Elsa just two and a half months later. *CPAE*, Vol. 8, Doc. 254; Fölsing, *Albert Einstein*, 425, 427

* Separation from Mileva was a matter of life and death for me. . . . Thus I deprive myself of my boys, whom I still love tenderly.

To Helene Savić, September 8, 1916. *CPAE*, Vol. 8,
Doc. 258

* I've been so preoccupied with what would happen
in the event of my death that I'm surprised to find
myself still alive.

To Mileva, April 23, 1918, after attending to legal
paperwork that would financially take care of her
and the boys in case of his death. *CPAE*, Vol. 8,
Doc. 515

* Mileva was absolutely insufferable when we were
together. When we are not, I can like her quite well;
she seems all right to me, even as the mother of my
boys.

To Michele Besso, July 29, 1918. *CPAE*, Vol. 8,
Doc. 591

She never reconciled herself to the separation and
divorce, and a disposition developed reminiscent of
the classical example of Medea. This darkened the
relations with my two boys, to whom I was attached with tenderness. This tragic aspect of my life
continued undiminished until my advanced age.

To Carl Seelig, May 5, 1952, about Mileva. Einstein
Archive 39-020

Einstein began a long-distance affair with his cousin Elsa, who lived in Berlin, in 1912, while he was still married to Mileva and living in Zurich. The affair continued after the family moved to Berlin in 1914. He was not divorced from Mileva, who soon returned to Zurich, until February 1919. In June of that year he married Elsa, though for many years he had been telling friends that he did not intend to marry her and had even considered marrying her daughter Ilse instead. At one time he had also had his eye on Paula, Elsa's younger sister. See various letters in *CPAE*, Vol. 8, and Stern, *Einstein's German World*, 105n

I will always destroy your letters, as is your wish. I have already destroyed the first one.

To Elsa, April 30, 1912, responding to her misgivings about their affair. *CPAE*, Vol. 5, Doc. 389

* I must love someone. Otherwise it is a miserable existence. And that someone is you.

Ibid.

I suffer even more than you because you suffer only for what you do *not* have.

To Elsa, May 7, 1912, alluding to his difficult wife, Mileva. *CPAE*, Vol. 5, Doc. 391

I am writing so late because I have misgivings about our affair. I have a feeling that it will not be good for us, nor for the others, if we form a closer attachment.

To Elsa, May 21, 1912. *CPAE*, Vol. 5, Doc. 399

I now have someone about whom I can think with unrestrained pleasure and for whom I can live. . . . We will have each other, something we have missed so terribly, and will give each other the gift of stability and an optimistic view of the world.

To Elsa, October 10, 1913. *CPAE*, Vol. 5, Doc. 476

If you were to recite for me the most beautiful poem . . . my pleasure would not even approach the pleasure I felt when I received the mushrooms and goose cracklings you prepared for me; . . . you will surely not despise the domestic side of me that is revealed by this disclosure.

To Elsa, November 7, 1913. *CPAE*, Vol. 5, Doc. 482

* I really delight in my local relatives, especially in a cousin of my age, with whom I am linked by an old friendship. It is mostly because of this that I am

Ilse Einstein, circa 1925. (Courtesy of Einstein Archive, Boston)

accustoming myself very well to the large city [Berlin], which is otherwise loathsome to me.

> To Paul Ehrenfest, ca. April 10, 1914, on his adjustment to life in Berlin. *CPAE*, Vol. 8, Doc. 2

Margot Einstein, circa 1930. (Courtesy of Einstein Archive, Boston)

*I would take only one of the women with me, either Elsa or Ilse. The latter is more suitable because she is healthier and more practical.

To Fritz Haber, October 7, 1920, on taking a traveling companion on a lecture trip to Norway. Einstein Archive 12-327. He neglected to mention

that he had also been infatuated with Ilse, Elsa's
daughter, before Elsa's and his marriage (see Ilse's
letter to Georg Nicolai, May 22, 1918, *CPAE*, Vol. 8,
Doc. 545).

ABOUT OR TO HIS CHILDREN

With Mileva, Einstein had two sons, Hans Albert
and Eduard, and a daughter, referred to as "Lieserl";
through his marriage to Elsa, he had two
stepdaughters, Ilse and Margot. Lieserl was born
in January 1902 before Einstein and Mileva were
married, and she was presumably given up for
adoption or died after the effects of scarlet fever; no
mention is made of her after September 1903, and
Einstein never saw her. See *CPAE*, Vol. 5, and *The
Love Letters*. Only Hans Albert had children. Eduard
developed schizophrenia at the age of twenty,
though up to that time he had been a somewhat
fragile but essentially healthy young man pursuing a
medical education. Eduard remained in Switzerland
all of his life, and Einstein's only contact with him
after leaving Europe in 1933 was through his
biographer, Carl Seelig; he told Seelig that he never
wrote to Eduard afterwards for reasons he could not
analyze himself. Einstein Archive 39-060

I'm very sorry about what has befallen Lieserl. It's
so easy to suffer lasting effects from scarlet fever. If

this will only pass. As what is the child registered? We must take precautions that problems don't arise for her later.

> This somewhat cryptic (to the reader) letter was sent to Mileva ca. September 19, 1903. Registering a child may indicate the parents' intention of giving it up. They may have considered Lieserl's illegitimacy a threat to Einstein's provisional federal appointment at the Swiss Patent Office. See *CPAE*, Vol. 5, Doc. 13, n. 4

* Nowhere else is it as nice for boys as in Zurich. . . . Boys aren't pestered too much with homework there, nor with the need to be too well dressed and well mannered.

> To son Hans Albert, after the boys returned to Zurich with their mother, January 25, 1915. *CPAE*, Vol. 8, Doc. 48

* At the time we [he and Mileva] were separating from each other, the thought of leaving the children stabbed me like a dagger every morning when I awoke, but I have never regretted the step in spite of it.

> To Heinrich Zangger, November 26, 1915. *CPAE*, Vol. 8, Doc. 152

* Today I'm sending off some toys for you and Tete. Don't neglect your piano, my Adu; you don't know how much pleasure you can give to others, as well

as to yourself, when you can play music nicely. . . . Another thing, brush your teeth every day, and if a tooth is not quite all right, go to the dentist immediately. I also do the same and am now very happy that I have healthy teeth. This is *very important*, as you will realize yourself later on.

> To Hans Albert ("Adu"; "Tete" is Eduard), ca. April 1915. *CPAE*, Vol. 8, Doc. 70. In a letter later that year he urges the two boys to take calcium chloride after every meal to promote strong tooth and bone development.

* I will try to be together with you for a month every year so that you will have a father who is close to you and can love you. You can also learn a lot of good things from me that no one else can offer you so easily. The things I have gained from so much strenuous work should be of value not only to strangers but especially to my own boys. In the last few days I completed one of the finest papers of my life. When you are older I'll tell you about it.

> To eleven-year-old Hans Albert, November 4, 1915, also referring to his paper on the general theory of relativity. *CPAE*, Vol. 8, Doc. 134

* Albert is now gradually entering the age at which I can mean very much to him. . . . My influence will be limited to the intellectual and esthetic. I want to teach him mainly to think, judge, and appreciate

things objectively. For this I need several weeks a year—a few days would only be a short thrill with no deeper value.

> To Mileva, who was afraid that her own
> relationship with Hans Albert would suffer if he
> had too much contact with his father, December 1,
> 1915. *CPAE*, Vol. 8, Doc. 159

* I am very glad that you enjoy the piano so much. I have one in my little apartment, too, and play it every day. I also play the violin a lot. Maybe you can practice something to accompany a violin, and then we can play at Easter when we are together.

> To Hans Albert, March 11, 1916. Einstein was not
> able to come at Christmas because of the difficulty
> of crossing borders during wartime, so he planned
> a trip at Easter. *CPAE*, Vol. 8, Doc. 199

* You still make so many writing errors. You must take care in that regard: it makes a bad impression when words are misspelled.

> To Hans Albert, March 16, 1916. *CPAE*, Vol. 8,
> Doc. 202

* My compliments on the good condition of our boys. They are in such excellent physical and emotional shape that I could not wish for more. And I know this is for the most part due to the proper upbringing you are providing.... They came to meet me spontaneously and sweetly.

> To Mileva during his visit to Zurich, April 8, 1916.
> *CPAE*, Vol. 8, Doc. 211

* I am writing you now for the third time without receiving a reply from you. Don't you remember your father anymore? Are we never going to see each other again?

> To Hans Albert, September 26, 1916. Einstein
> learned that the boys had become angry at him.
> They reconciled and continued to write
> occasionally, while Einstein visited about once a
> year during wartime. *CPAE*, Vol. 8, Doc. 261; see
> also Doc. 258

* You don't need to worry about grades. Just see to it that you keep up with the work and that you don't have to repeat a year. It is not necessary to have good marks in everything.

> To Hans Albert, October 13, 1916. *CPAE*, Vol. 8,
> Doc. 263

* Although I am over here, you do have a father who loves you more than anything else and who is constantly thinking of you and caring about you.

> Ibid., regarding their separation

* I could be a grandpa now too if my [Hans] Albert hadn't married such a *Schlemilde*.

> To his uncle, Caesar Koch, who had just become
> a grandfather, ca. 1927–1928. Einstein had
> vigorously opposed Hans Albert's marriage to
> Frieda Knecht, nine years his senior; but the couple
> remained together until Frieda's death. By 1930
> they had made Einstein a grandpa, too, with the
> birth of Bernhard. See Sotheby's catalog, June 26,
> 1998, 424

It is a thousand pities for the boy that he must pass his life without the hope of a normal existence. Since the insulin injections have proved unsuccessful, I have no further hopes from the medical side. I think it is better on the whole to let Nature run its course.

> To Michele Besso, November 11, 1940, about son
> Eduard. Einstein Archive 7-378

* There is a block behind it that I cannot fully analyze. But one factor is that I think I would arouse painful feelings of various kinds in him if I made an appearance in whatever form.

> To Carl Seelig, January 4, 1954, stating why he was
> not in touch with Eduard. In his will, Einstein left
> a larger amount of money to Eduard than to Hans
> Albert.

Eduard Einstein, circa 1955. (Courtesy of Leo Baeck Institute, New York)

It is a joy for me to have a son who has inherited the chief trait of my personality: the ability to rise above mere existence by sacrificing onself through the years for an impersonal goal. This is the best, indeed the only way in which we can make ourselves independent from personal fate and from other human beings.

To Hans Albert, May 1, 1954. Quoted in Highfield and Carter, *The Private Lives*, 258

Unfortunately, I have to admit that I didn't think about it, but your wife reminded me.

> To Hans Albert on his fiftieth birthday, May 1954.
> Quoted in an interview with Bernard Mayer, in
> Whitrow, *Einstein*, 21

When Margot speaks, you see flowers growing.

> Commenting on his stepdaughter Margot's love of
> nature. Quoted by friend Frieda Bucky in "You
> Have to Ask Forgiveness," *The Jewish Quarterly* 15,
> no. 4 (Winter 1967–68), 33

ABOUT HIS SISTER, MAJA,
AND MOTHER, PAULINE

Yes, but where are its wheels?

> Two-and-a-half-year-old Albert, after the birth of
> Maja in 1881, upon being told he would now have
> something to play with. In "Biographical Sketch,"
> by Maja Winteler-Einstein, *CPAE*, Vol. 1, lvii

My mother and sister seem somewhat petty and philistine to me, despite the sympathy I feel for them. It is interesting how life gradually changes us in the very subtleties of our soul, so that even the

closest of family ties dwindle into habitual friendship. Deep inside we no longer understand one another and are incapable of empathizing with the
other, or know what emotions move the other.

> To Mileva Marić, early August 1899. *The Love
> Letters*, 9; *CPAE*, Vol. 1, Doc. 50

My mother has died. . . . We are all completely ex-
hausted. One feels in one's bones the significance of
blood ties.

> To Heinrich Zangger, early March 1920. Einstein
> Archive 39-732

I know what it's like to see one's mother go
through the agony of death and be unable to help;
there is no consolation. We all have to bear such
heavy burdens, for they are unalterably linked to
life.

> To Hedwig Born, June 18, 1920. In Born, *Born-
> Einstein Letters*, 29

Albert and Elsa Einstein, going to America aboard the SS *Rotterdam*, 1921. (Courtesy of AIP Emilio Segrè Visual Archives)

I am happy to be in Boston. I have heard of Boston as one of the most famous cities in the world and the center of education. I am happy to be here and expect to enjoy my visit to this city and to Harvard.

On his visit to the city with Chaim Weizmann. *New York Times*, May 17, 1921. Contributed by A. J. Kox in response to the many quotations about Princeton in this book (see below).

The smile on the faces of the people . . . is symbolic of one of the greatest assets of the American. He is friendly, self-confident, optimistic—and not jealous.

From an interview, *Nieuwe Rotterdamsche Courant*, 1921; also quoted in *Berliner Tageblatt*, July 7, 1921; reprinted in *Ideas and Opinions*, 3–7

The American lives even more for his goals, for the future, than the European. Life for him is always becoming, never being. . . . He is less of an individualist than the European . . . more emphasis is put on the "we" than the "I."

Ibid.

I have warm admiration for American institutes of scientific research. We are unjust in attempting to ascribe the increasing superiority of American research work exclusively to superior wealth; devo-

tion, patience, a spirit of comradeship, and a talent
for cooperation play an important part in its success.

Ibid.

* American men, though they are hardworking, are
nothing more than toy dogs of the women, who
like to spend money ... and wrap themselves in a
veil of excess.

Quoted in the *New York Times*, July 8, 1921. This
statement outraged many Americans, and they did
not forget it or forgive Einstein for a long time.

Even if Americans are less scholarly than Germans,
they do have more enthusiasm and energy, causing a wider dissemination of new ideas among the
people.

Quoted in the *New York Times*, July 12, 1921

A firm approach is indispensable everywhere in
America; otherwise one receives no payment and
little esteem.

To Maurice Solovine, January 14, 1922. Einstein
Archive 21-157; published in *Letters to Solovine*, 49

Never have I experienced from the fair sex such an
energetic rejection of all my advances; if it *has* happened, it was never from so many at once.

Answer to an American women's organization,
January 4, 1928, which had protested Einstein's

visit to America on political grounds. Einstein
Archive 48-818; published in *Ideas and Opinions*, 7

* Here in Pasadena it is like paradise.... Always
sunshine and fresh air, gardens with palm and pep-
per trees, and friendly people who smile at one and
ask for autographs.

> To the Lebach family during the days before smog,
> January 16, 1931, on the city in which the
> California Institute of Technology is located.
> Einstein Archive 47-373

* [America], this land of contrasts and surprises,
which leaves one filled alternately with admiration
and incredulity. One feels more attached to the Old
Europe, with its heartaches and hardships, and is
glad to return there.

> To Queen Elizabeth of Belgium, February 9, 1931,
> revealing a touch of homesickness during his
> three-month stay in America. Einstein Archive
> 32-349

* For the long term I would prefer being in Holland
rather than in America.... Besides having a handful of really fine scholars, it is a boring and barren
society that would soon make you tremble.

> To Paul Ehrenfest, April 3, 1932, after his return to
> Europe. Einstein Archive 10-227

* If there were no newspapers here, I would live as on a newly discovered planet. People here regard Europe as something between a theater and a zoological garden.

> To Maurice Lecat, August 11, 1934

America is today the hope of all honorable men who respect the rights of their fellow men and who believe in the principles of freedom and justice.

> "Message for Germany," dictated over the telephone on December 7, 1941, the day that Pearl Harbor was bombed, to a White House correspondent. Quoted in Nathan and Norden, *Einstein on Peace*, 320

The only justifiable purpose of political institutions is to assure the unhindered development of the individual. . . . That is why I consider myself to be particularly fortunate to be an American.

> Ibid.

There is, however, a somber point in the social outlook of Americans. Their sense of equality and human dignity is mainly limited to people of white skin. . . . The more I feel like an American, the more this situation pains me.

> From an address at Lincoln University, Pennsylvania, a university for black men, upon receipt of an honorary doctorate, May 3, 1946.

Einstein supported the fledgling civil rights movement, perhaps influenced by Paul Robeson, a black opera singer, former athlete, and early civil rights advocate, who had been born in Princeton. Quoted in *Out of My Later Years*, under "The Negro Question," 127; see also "Blacks/Racism/Slavery" under "Miscellaneous Subjects"

The separation [between Jews and Gentiles] is even more pronounced [in America] than it ever was anywhere in Western Europe, including Germany.

To Hans Muehsam, March 24, 1948. Einstein Archive 38-371

I hardly ever felt as alienated from people as I do right now. . . . The worst is that nowhere is there anything with which one can identify. Brutality and lies are everywhere.

To Gertrud Warschauer, July 15, 1950, about the McCarthy era. Einstein Archive 39-505

The German calamity of years ago repeats itself: people acquiesce without resistance and align themselves with the forces of evil.

To Queen Elizabeth of Belgium, January 6, 1951, about McCarthyism in America. Einstein Archive 32-400; also quoted in Nathan and Norden, *Einstein on Peace*, 554

I have become a kind of *enfant terrible* in my new homeland because of my inability to keep silent and swallow everything that happens here.

> To Queen Elizabeth of Belgium, March 28, 1954.
> Einstein Archive 32-410

ON HIS ADOPTED
HOMETOWN OF PRINCETON,
NEW JERSEY

I found Princeton fine. A pipe as yet unsmoked. Young and fresh.

> *New York Times*, July 8, 1921, reporting on his
> lecture trip to his future hometown

Princeton is a wondrous little spot, a quaint and ceremonious village of puny demigods on stilts. Yet, by ignoring certain social conventions, I have been able to create for myself an atmosphere conducive to study and free from distraction.

> To Queen Elizabeth of Belgium, November 20,
> 1933. Einstein Archive 32-369

My fame begins outside of Princeton. My word counts for little in Fine Hall.

> On his lack of decision-making power on the Princeton campus, 1934–1940. The old Fine Hall is now Jones Hall, where the East Asian Studies department is located. Quoted in Infeld, *Quest*, 302

I am very happy with my new home in friendly America and in the liberal atmosphere of Princeton.

> From an interview, *Survey Graphic* 24 (August 1935), 384, 413

As an elderly man, I have remained estranged from the society here.

> To Queen Elizabeth of Belgium, February 16, 1935. Einstein Archive 32-385

I am privileged by fate to live here in Princeton as if on an island that . . . resembles the charming palace garden in Laeken [Belgium]. Into this small university town the chaotic voices of human strife barely penetrate. I am almost ashamed to be living in such a place while all the rest struggle and suffer.

> To Queen Elizabeth of Belgium, March 20, 1936. Einstein Archive 32-387

In the face of all the heavy burdens I have borne in recent years, I feel doubly thankful that there has fallen on my lot in Princeton University a place for work and a scientific atmosphere which could not be better or more harmonious.

> To university president Harold Dodds, January 14, 1937. At the time, Einstein's office was temporarily located on the Princeton campus even though he was a member of the Institute for Advanced Study, a separate institution whose campus had not yet been built. Einstein Archive 52-823

A banishment to paradise.

> On going to Princeton. Quoted in Sayen, *Einstein in America*, 64

You are surprised, aren't you, at the contrast between my fame throughout the world . . . and the isolation and quiet in which I live here. I wished for this isolation all my life, and now I have finally achieved it here in Princeton.

> Quoted in Frank, *Einstein: His Life and Times*, 297

On Death

STATE OF NEW JERSEY   NO. 227

OFFICE OF REGISTRAR OF VITAL STATISTICS

of   **Princeton Borough, Mercer County**

City, Borough or Township and County

This is to Certify that the following is correctly copied from a record of Death in my office.

NAME OF DECEASED			PLACE OF DEATH		DATE OF DEATH		
Albert Einstein			Princeton Hospital		April 18, 1955		
SOCIAL SECURITY NUMBER	SEX	COLOR	MARITAL CONDITION	DATE OF BIRTH		AGE	
					YRS	MOS	DAYS
	Male	White	Widower	March 14, 1879	76	1	4
PLACE OF BIRTH			CAUSE OF DEATH				
Ulm, Germany			Rupture of Arteriosclerotic Aneurysm of Abdominal Aorta.				
			SUPPLEMENTAL INFORMATION IF DEATH WAS DUE TO EXTERNAL CAUSES				
ACCIDENT, SUICIDE OR HOMICIDE			SPECIFY	DATE OF OCCURRENCE			
WHERE DID INJURY OCCUR?							
		CITY OR TOWN		COUNTY		STATE	
DID INJURY OCCUR IN OR ABOUT HOME, ON FARM, IN INDUSTRIAL PLACE, IN PUBLIC PLACE?					SPECIFY TYPE OF PLACE		
WHILE AT WORK?			MEANS OF INJURY				
NAME OF PERSON WHO CERTIFIED CAUSE OF DEATH				ADDRESS			
Guy K. Dean, Jr., M. D.				Plainsboro, N. J.			

David T. Blake

Registrar of Vital Statistics

Borough Hall, Princeton, N. J.

Address

April 26, 19 55

Date of Issue

I have firmly resolved to bite the dust, when my time comes, with a minimum of medical assistance, and up to then I will sin to my wicked heart's content.

To Elsa Einstein, August 11, 1913. *CPAE*, Vol. 5, Doc. 466

* The old who have died live on in the young ones. Don't you feel this now as you mourn, when you look at your children?

To Hedwig Born, April 18, 1920, after the death of her mother. In Born, *Born-Einstein Letters*, 29

I feel myself so much a part of everything living that I am not the least concerned with the beginning or ending of the concrete existence of any one person in this eternal flow.

Ibid.

Our death is not an end if we can live on in our children and the younger generation. For they are us; our bodies are only wilted leaves on the tree of life.

To the widow of Dutch physicist Heike Kamerlingh-Onnes, February 25, 1926. Einstein Archive 14-389

Neither on my deathbed nor before will I ask my-self such a question. Nature is not an engineer or a contractor, and I myself am a part of Nature.

> In answer to a question concerning what facts
> would determine if his life was a success or failure,
> November 12, 1930. Einstein Archive 45-751; also
> quoted in Dukas and Hoffmann, *Albert Einstein, the
> Human Side*, 92

I feel unable to participate in your projected TV broadcast "The Last Two Minutes." It seems to me not so relevant how people are to spend the last two minutes before their final deliverance.

> In answer to a request that he participate in a
> television program on how some famous people
> would spend the last two minutes of their lives,
> August 26, 1950. Einstein Archive 60-684

I myself should also be dead already, but I am still here.

> To E. Schaerer-Meyer, July 27, 1951. Einstein
> Archive 60-525

Look deep, deep into nature, and then you will un-derstand everything better.

> To Margot Einstein, after the death of his sister,
> Maja, 1951. Quoted by friend Hanna Loewy in
> A&E Television's Einstein biography, VPI
> International, 1991

* Brief is this existence, like a brief visit in a strange house. The path to be pursued is poorly lit by a flickering consciousness whose center is the limiting and separating "I." ... When a group of individuals becomes a "we," a harmonious whole, they have reached as high as humans can reach.

> At the graveside of physicist Rudolf Ladenburg, April 1954. See Stern, *Einstein's German World*, 163

To one bent on age, death will come as a release. I feel this quite strongly now that I have grown old myself and have come to regard death like an old debt, at long last to be discharged. Still, instinctively one does everything possible to postpone the final settlement. Such is the game that Nature plays with us.

> To Gertrud Warschauer, February 5, 1955. Quoted in Nathan and Norden, *Einstein on Peace*, 616

I want to go when *I* want. It is tasteless to prolong life artificially. I have done my share; it is time to go. I will do it elegantly.

> Quoted by Helen Dukas in her letter to Abraham Pais, an Einstein biographer, April 30, 1955. See Pais, *Subtle Is the Lord*, 477; also quoted in A&E Television's Einstein biography, VPI International, 1991

I want to be cremated so people won't come to worship at my bones.

> Quoted by Einstein biographer Abraham Pais,
> *Manchester Guardian*, December 17, 1994

* My house will certainly not become a place of pilgrimage, where pilgrims can come to view the bones of the saint.

> In reply to a student's question on what would
> become of his house after his death. Recalled by
> John Wheeler in French, *Einstein*, 22

On Education and
Academic Freedom

Einstein receiving an honorary degree. (Courtesy of
Leo Baeck Institute, New York)

The inclination of the pupil for a particular profession must not be neglected, especially because such inclination usually asserts itself at an early age, being occasioned by personal gifts, by example of other members of the family, and by various other circumstances.

1920. In Moszkowski, *Conversations with Einstein*, 65

Most teachers waste their time by asking questions that are intended to discover what a pupil does *not* know, whereas the true art of questioning is to discover what the pupil *does* know or is capable of knowing.

Ibid.

It is not so very important for a person to learn facts. For that he does not really need college. He can learn them from books. The value of an education in a liberal arts college is not the learning of many facts, but the training of the mind to think something that cannot be learned from textbooks.

1921, on Thomas Edison's opinion that a college education is useless. Quoted in Frank, *Einstein: His Life and Times*, 185

It would be better if you began to teach others only after you yourself have learned something.

> To Arthur Cohen, age 12, who had submitted a
> paper to Einstein, December 26, 1928. Einstein
> Archive 25-044. Cohen's sister-in-law, Betty,
> contacted me after reading this quotation. Young
> Arthur eventually went to Stanford, then received
> a Ph.D. in botany from Harvard, and became a
> professor at Washington State University. It seems
> that he heeded Einstein's advice.

Never regard your study as a duty, but as the enviable opportunity to learn the liberating influence of beauty for your own personal joy and for the profit of the community to which your later work will belong.

> In the Princeton University freshman publication,
> *The Dink*, December 1933

In the teaching of geography and history, a sympathetic understanding [should] be fostered for the characteristics of the different peoples of the world, especially for those whom we are in the habit of describing as "primitive."

> From "The Schools and the Problem of Peace," an
> address delivered at the Conference of the
> Progressive Education Association, November 23,
> 1934. In *Einstein on Humanism*, 71–72

Humiliation and mental oppression by ignorant and selfish teachers wreak havoc in the youthful mind that can never be undone and often exert a baleful influence in later life.

In *Almanak van het Leidsche Studentencorps* (Leiden: Doesburg-Verlag, 1934)

To me the worst thing seems to be for a school principally to work with the methods of fear, force, and artificial authority. Such treatment destroys the sound sentiments, the sincerity, and the self-confidence of the pupil.

Address at a convention at the State University of New York in Albany, October 15, 1936. In *School and Society* 44 (1936), 589–592. See also the *New York Times*, October 16, 1936, 11:1; published as "On Education," in *Out of My Later Years*, 36

The aim [of education] must be the training of independently acting and thinking individuals who, however, see in the service to the community their highest life achievement.

Ibid., 35

The school should always have as its aim that the young person leave it as a harmonious personality, not as a specialist.

Ibid., 39

Otherwise, he—with his specialized knowledge—more closely resembles a well-trained dog than a harmoniously developed person.

New York Times, October 5, 1952

Freedom of teaching and of opinion in book or press is the foundation for the sound and natural development of any people.

At a gathering for Freedom of Opinion, 1936.
Quoted in *Einstein on Humanism*, 50

* Whereas life is unplanned and chaotic, the educational system operates according to a definite scheme. . . . That explains . . . why education is such an important political instrument: there is always the danger that it may become an object of exploitation by contending political groups.

Message to the New Jersey Education Association, Atlantic City, 1939. See the *New York Times*, November 11, 1939, 34:2

The crippling of individuals I consider the worst evil of capitalism. Our whole educational system suffers from this evil. An exaggerated competitive attitude is inculcated into the student, who is trained to worship material success as a preparation for his future career.

From "Why Socialism?" *Monthly Review*, May 1949

Teaching should be such that what is offered is perceived as a valuable gift and not as a hard duty.

New York Times, October 5, 1952

By academic freedom I understand the right to search for truth and to publish and teach what one holds to be true. This right also implies a duty: one must not conceal any part of what one has recognized to be true. It is evident that any restriction of academic freedom acts in such a way as to hamper the dissemination of knowledge among the people and thereby impedes national judgment and action.

Statement for a conference of the Emergency Civil Liberties Committee, March 13, 1954. Quoted in Nathan and Norden, *Einstein on Peace*, 551

The Einstein family, *left*, with Rabindranath Tagore, Indian poet and philosopher, and his family, Berlin, 1931. (Courtesy of Leo Baeck Institute, New York)

Now he has departed from this strange world a little ahead of me. That signifies nothing. For us believing [*gläubige*] physicists, the distinction between past, present, and future is only a stubbornly persistent illusion.

> On lifelong friend Michele Besso, in a letter of
> condolence to the Besso family, March 21, 1955,
> less than a month before his own death. Einstein
> Archive 7-245

What I admired most in him as a human being is that he managed to live for so many years not only in peace but also in lasting harmony with a woman—an undertaking in which I twice failed rather miserably.

> Ibid.

ON NIELS BOHR

Not often in my life has a person given me such joy by his presence as you have.

> To Niels Bohr, May 2, 1920. Einstein Archive 8-065

Bohr was here, and I'm as enamored of him as you are. He is like an extremely sensitive child who moves around in this world in a sort of trance.

To Paul Ehrenfest, May 4, 1920. Einstein Archive
9-486

He is truly a man of genius.... I have full confidence in his way of thinking.

To Paul Ehrenfest, March 23, 1922. Einstein
Archive 10-035

* What is so marvelously attractive about Bohr as a scientific thinker is his rare blend of boldness and caution; seldom has anyone possessed such an intuitive grasp of hidden things combined with such a strong critical sense.... He is unquestionably one of the greatest discoverers of our age in the scientific field.

"Niels Bohr," in Einstein, *Essays in Science* (1934),
47

He utters his opinions like one who perpetually casts about, and never like one who believes he holds the whole defining truth.

To Bill Becker, March 20, 1954. Einstein Archive
8-109

ON LOUIS BRANDEIS

I know of no other person who combines such profound intellectual gifts with such self-renunciation while finding the whole meaning of his life in quiet service to the community.

To Supreme Court Justice Louis Brandeis,
November 10, 1936. Einstein Archive 35-046

ON CHARLIE CHAPLIN

* [He] had set up a Japanese theater in his home, with authentic Japanese dances being performed by Japanese girls. Just as in his films, Chaplin is an enchanting person.

To the Lebach family, January 16, 1931, after
visiting Chaplin in Hollywood. (Later in the
month, Einstein attended the premiere of
Chaplin's *City Lights* with him.) Einstein Archive
47-373

ON MARIE CURIE

I do not believe Mme Curie is power-hungry or hungry for whatever. She is an unpretentious, honest person with more than her share of responsibilities and burdens. She has a sparkling intelligence, but despite her passionate nature she is not attractive enough to present a danger to anyone.

To Heinrich Zangger, November 7, 1911,
regarding Curie's alleged affair with married
French physicist Paul Langevin. *CPAE*, Vol. 5,
Doc. 303

* I am compelled to tell you how much I have come to admire your intellect, your vitality, and your honesty, and that I consider myself fortunate to have made your personal acquaintance in Brussels.

To Marie Curie, November 23, 1911. *CPAE*, Vol. 8,
Doc. 312a on p. 7

I am deeply grateful to you and your friends that you so cordially allowed me to participate in your daily life. To witness such marvelous camaraderie among such people is the most uplifting thing I can think of. Everything looked so natural and uncomplicated with you, like a good work of art. . . . I wish to ask your forgiveness if by any chance my crude manners sometimes made you feel uncomfortable.

To Marie Curie, April 3, 1913. *CPAE*, Vol. 5,
Doc. 435

Madame Curie is very intelligent but as cold as a herring, meaning that she lacks all feelings of joy and sorrow. Almost the only way she expresses her feelings is to rail against things she doesn't like. And she has a daughter who is even worse—like a grenadier. This daughter is also very gifted.

To Elsa Löwenthal, ca. August 11, 1913. *CPAE*, Vol. 5, Doc. 465

Her strength, her purity of will, her austerity toward herself, her objectivity, her incorruptible judgment—all these were of a kind seldom found in a single individual.... Once she had recognized a certain way as a right one, she pursued it without compromise and with extreme tenacity.

At Curie memorial celebration, Roerich Museum, New York, November 23, 1934. Einstein Archive 5-142

ON PAUL EHRENFEST

His sense of inadequacy, objectively unjustified, plagued him incessantly, often robbing him of the peace of mind necessary for tranquil research.... His tragedy lay precisely in an almost morbid lack of self-confidence.... The strongest relationship in his life was that toward his wife and fellow worker ... his intellectual equal.... He repaid her with a veneration and love such as I have not often witnessed in my life.

After physicist and close friend Paul Ehrenfest's suicide. Ehrenfest had shot and killed his ill sixteen-year-old son first. Quoted in *Almanak van het Leidsche Studentencorps* (Leiden: Doesburg-Verlag, 1934)

ON MICHAEL FARADAY

This man loved mysterious Nature as a lover loves his distant beloved.

To Gertrud Warschauer, December 27, 1952.
Einstein Archive 39-517

ON OR TO SIGMUND FREUD

* Why do you emphasize happiness in my case? You, who have gotten under the skin of so many people and, indeed, of humanity, have had no occasion to slip under mine.

To Sigmund Freud, March 22, 1929, in reply to Freud's letter on Einstein's fiftieth birthday, in which he congratulated him for being a "happy one." Einstein Archive 32-530

The old one ... had a sharp vision; no illusions lulled him to sleep except for an often exaggerated faith in his own ideas.

To A. Bacharach, July 25, 1949. Einstein Archive 57-629

ON GALILEO

Alas, you find [vanity] in so many scientists! It has always pained me that Galileo did not acknowledge the work of Kepler.

To I. Bernard Cohen, April 1955, in an interview shortly before his death. *Scientific American*, July 1955, 69

ON GANDHI

* I admire Gandhi greatly but I believe there are two weaknesses in his program. Nonresistance is the most intelligent way to face difficulty, but it can be practiced only under ideal conditions.... It could not be carried out against the Nazi party today. Then, Gandhi makes a mistake in trying to abolish the machine from modern civilization. It is here, and it must be dealt with.

From an interview, *Survey Graphic* 24 (August 1935), 384, 413

A leader of his people, unsupported by any outward authority: a politician whose success rests not upon craft or the mastery of technical devices, but simply on the convincing power of his personality; a victorious fighter who has always scorned the use of force; a man of wisdom and humility, armed with resolve and inflexible consistency, who has devoted all his strength to the uplifting of his people and the betterment of their lot; a man who has confronted the brutality of Europe with the dignity of the simple human being, and thus at all times risen superior.

Generations to come, it may well be, will scarce believe that such a one as this ever in flesh and blood walked upon this earth.

> Statement on the occasion of Gandhi's seventieth birthday, 1939. In *Einstein on Humanism*, 94

I believe that Gandhi's views were the most enlightened among all of the political men of our time. We should strive to do things in his spirit; not to use violence in fighting for our cause, but by nonparticipation in what we believe is evil.

> From a United Nations Radio interview, June 16, 1950, recorded in the study of Einstein's home in Princeton. Reprinted in the *New York Times*, June 19, 1950; also quoted in Pais, *Einstein Lived Here*, 110

Gandhi, the greatest political genius of our time . . . gave proof of what sacrifice man is capable once he has discovered the right path.

> To Asian Congress for World Federation, November 1952. Quoted in Nathan and Norden, *Einstein on Peace*, 584

Gandhi's development resulted from extraordinary intellectual and moral forces in combination with political ingenuity and a unique situation.

1953. In ibid., 594

ON GOETHE

I feel in him a certain condescending attitude toward the reader, a certain lack of humility that, especially when it comes from great men, is comforting [to the reader].

To L. Caspar, April 9, 1932. Einstein Archive 49-380

* I admire Goethe as a poet without peer, and as one of the smartest and wisest men of all time. Even his scholarly ideas deserve to be held in high esteem, and his faults are those of any great man.

Ibid.

ON HITLER

In Hitler we have a man with limited intellectual abilities, unfit for any useful work, bursting with envy and bitterness against all of those whom circumstance and nature had favored over him. . . . He picked up human flotsam on the street and in the taverns and organized them around himself. That's how he became a politician.

From an unpublished manuscript, 1935. Quoted in Nathan and Norden, *Einstein on Peace*, 263–264

ON HEIKE KAMERLINGH-ONNES

A life has ended that will always remain a role model for future generations. . . . No other person have I known for whom duty and joy were one and the same. This was the reason for his harmonious life.

To the Dutch physicist's widow, February 25, 1926. Einstein Archive 14-389

ON KANT

What seems to me the most important thing in Kant's philosophy is that it speaks of a priori concepts for the construction of science.

At a discussion in the Société Française de Philosophie, July 1922. Quoted in *Bulletin Société Française de Philosophie* 22 (1922), 91; reprinted in *Nature* 112 (1923), 253

*If Kant had known what is known to us today of the natural order, I am certain that he would have fundamentally revised his philosophical conclusions. Kant built his structure upon the foundations of the world outlook of Kepler and Newton. Now that the foundation has been undermined, the structure no longer stands.

From an interview with Chaim Tschernowitz, *The Jewish Sentinel*, September 1931, 19, 44, 50

* Kant, thoroughly convinced of the indispensability of certain concepts, took them—just as they are selected—to be necessary premises for every kind of thinking and differentiated them from concepts of empirical origin.

"Autobiographical Notes," in Schilpp, *Albert Einstein: Philosopher-Scientist*, 13

ON KEPLER

Kepler was one of the few who are simply incapable of doing anything but stand up openly for their convictions in every field. . . . [His] lifework was possible only when he succeeded in freeing himself to a great extent from the intellectual traditions into which he was born. . . . He does not speak of it, but the inner struggle is reflected in his letters.

From preface to *Johannes Kepler: Life and Letters*, ed. Carola Baumgardt (New York: Philosophical Library, 1951)

TO LOVER AND ALLEGED SOVIET SPY, MARGARITA KONENKOVA

In 1998, a group of letters written by Einstein was put up for auction at Sotheby's in New York. They were written to a woman with whom he had had a love affair before and during World War II while living in Princeton. The woman was Margarita Konenkova, and, according to a book published in 1995 by former Soviet spymaster Pavel Sudoplatov, she was a Russian agent whose official mission was to introduce Einstein to the Soviet vice-consul in New York and "to influence Oppenheimer and other prominent American scientists whom she frequently met in Princeton." She did succeed in introducing Einstein to the Soviet vice-consul, Pavel Mikhailev, and Einstein refers to him in their letters. But Sudoplatov's account otherwise becomes suspect because, among other things, he places Oppenheimer in Princeton at the time, while he was actually 2,000 miles away in Los Alamos, New Mexico, helping to design the bomb; he did not come to Princeton until 1947, two years after Mrs. Konenkova had left. Einstein was sixty-six at the time the letters started in late 1945, and Mrs. Konenkova, whose husband Sergei created the bronze bust of Einstein at the Institute for Advanced Study in 1935, was in her mid-forties (though the *New York Times* put her age at 51). She and Sergei were Russian emigrés who lived in Greenwich Village from the early 1920s to 1945, when they were recalled to the Soviet Union. Margarita was a good friend of Margot Einstein's, whose ex-husband had been an attaché at the Soviet Embassy in Berlin in the early 1930s. She was known to have had affairs with other prominent men, and there is no indication that Einstein was aware that she might be a spy. See the *New York Times*, June 1, 1998, A1, and Sotheby's catalog, June 26, 1998; all the letters below are in Sotheby's catalog or in the *New York Times*, slightly differently translated.

* Because of your great love for your homeland, you would probably have become bitter in time if you had not taken this step [to return to Russia]. For unlike me, you have decades of active work and life ahead of you, whereas for me everything indicates . . . that my days will have run their course before too long. I think of you often.

Letter of November 8, 1945

* I recently washed my hair by myself, but not with great success; I am not as careful as you are. But everything here reminds me of you: . . . the dictionaries, the wonderful pipe which we thought was lost, and all the other little things in my hermit's cell; and also my lonely nest.

Letter of November 27, 1945. It appears that women liked to play with Einstein's hair. Another woman friend in Princeton is known to have cut his hair (obviously not frequently enough).

* People are living now [after the war] just as they were before . . . and it is clear that they have learned nothing from the horrors they have had to deal with. The little intrigues with which they had complicated their lives before are again taking up most of their thoughts. What a strange species we are.

Letter of December 30, 1945

* With best wishes and kisses, if this letter reaches you. And may the devil take anyone who intercepts it.

Letter of February 8, 1946

* I can imagine that the May Day ceremonies [in Moscow] must have been marvelous. But you know that I watch these exaggerated patriotic exhibitions with concern. I always try to convince people of the importance of cosmopolitan, reasonable, and fair thinking.

Letter of June 1, 1946

ON PAUL LANGEVIN

If he loves Mme Curie and she loves him, they do not have to run off together, because they have plenty of opportunities to meet in Paris. But I don't have the impression that anything special is going on between the two of them; rather, I found all three of them bound by a pleasant and innocent relationship.

To Heinrich Zangger, November 7, 1911, on the married French physicist's rumored affair with Marie Curie. The third person Einstein is referring to was Jean Perrin, another colleague. *CPAE*, Vol. 5, Doc. 303

There are so few in any generation in whom clear insight into the nature of things is joined with an intense feeling for the challenge of true humanity and the capacity for militant action. When such a man departs, he leaves a gap that seems unbearable to his survivors. . . . His desire to promote a happier life for all men was perhaps even stronger than his craving for pure intellectual enlightenment. No one who appealed to his social conscience ever went away empty-handed.

Quoted in *La Pensée*, February–March 1947

I had already heard of Langevin's death. He was one of my dearest acquaintances, a true saint, and talented besides. It is true that the politicians exploited his goodness, for he was unable to see through the base motives that were so foreign to his nature.

To Maurice Solovine, April 9, 1947. Einstein Archive 21-250; published in *Letters to Solovine*, 99

ON PHILIP LENARD

* I admire Lenard as a master of experimental physics; but he has not yet produced anything outstanding in theoretical physics, and his objections to the general theory of relativity are of such superficiality that up to now I did not think it necessary to answer to them in detail.

> On the staunch Nazi and anti-Semite. *Berliner Tageblatt*, August 27, 1920, 1–2. See also *CPAE*, Vol. 7, forthcoming

ON LENIN AND ENGELS

* I respect Lenin as a man who gave all his energy, at a total sacrifice of his personal life, to dedicating himself to the realization of socialist justice. I don't consider his methods appropriate. But one thing is certain: men such as he are the guardians and renewers of mankind's conscience.

> Statement for the League of Human Rights on Lenin's death, January 6, 1929. Einstein Archive 34-439

Outside Russia, Lenin and Engels are of course not valued as scientific thinkers and no one would be interested in refuting them as such. The same might also be the case in Russia, except there one doesn't dare say so.

To K. R. Leistner, September 8, 1932. Einstein
Archive 50-877

ON H. A. LORENTZ

Lorentz is a marvel of intelligence and exquisite tact. A living work of art! In my opinion he was the most intelligent of the theorists present [at the Solvay Congress in Brussels].

To Heinrich Zangger, November 1911, on the
Dutch physicist, whom Einstein loved and
admired. *CPAE*, Vol. 5, Doc. 305

My feeling of intellectual inferiority with regard to you cannot spoil the great delight of [our] conversation, especially because the fatherly kindness you show to all people does not allow any feeling of despondency to arise.

To Lorentz, February 18, 1912. *CPAE*, Vol. 5,
Doc. 360

He shaped his life like a precious work of art down to the smallest detail. His never-failing kindness and generosity and his sense of justice, coupled with a sure and intuitive understanding of people and human affairs, made him a leader in any sphere he entered.

Address at the grave of Lorentz, 1928. Published in *Ideas and Opinions*, 73

* To me personally, he meant more than all the others encountered in my lifetime.

Written in 1953 for the enlarged edition of *Mein Weltbild*, 31

People do not realize how great was the influence of Lorentz on the development of physics. We cannot imagine how it would have gone had Lorentz not made so many great contributions.

Quoted by Robert Shankland in French, *Einstein: A Centenary Volume*, 39

ON ERNST MACH

In him, the immediate pleasure gained in seeing and comprehending—Spinoza's *amor dei intellectualis*—was so strong that he looked at the world with the curious eyes of a child until well into old age, so that he could find joy and contentment in understanding how everything is connected.

Obituary for the philosopher whose critique of Newton played a role in Einstein's development of relativity theory, even though Mach himself was

critical of the theory. In *Physikalische Zeitschrift*,
April 1, 1916; *CPAE*, Vol. 6, Doc. 29

Mach was as good a scholar of mechanics as he was
a deplorable philosopher.

Quoted in *Bulletin Société Française de Philosophie* 22
(1922), 91; reprinted in *Nature* 112 (1923), 253; see
also *CPAE*, Vol. 6, Doc. 29, n. 6

ON ALBERT A. MICHELSON

I always think of Michelson as the artist in science.
His greatest joy seemed to come from the beauty of
the experiment itself and the elegance of the
method employed.

To Robert Shankland, September 17, 1953, on
the physicist who, with Edward Morley in 1881,
had already experimentally validated Einstein's
postulation that the speed of light is independent
of the frame of reference in which it is measured.
Einstein said that he was unaware of the
experiment when he wrote his 1905 paper on
the special theory of relativity. See discussion in
Fölsing, *Albert Einstein*, 217–219

ON NEWTON

His clear and wide-ranging ideas will retain their
unique significance for all time as the foundation of

our whole modern conceptual structure in the sphere of natural philosophy.

Statement for *The Times* (London), November 28, 1919

In my opinion, the greatest creative geniuses are Galileo and Newton, whom I regard in a certain sense as forming a unity. And in this unity Newton is [the one] who has achieved the most imposing feat in the realm of science.

1920. In Moszkowski, *Conversations with Einstein*, 40

In one person he combined the experimenter, the theorist, the mechanic, and, not the least, the artist of exposition.

From the "Introduction" to Newton, *Opticks* (McGraw-Hill, 1932)

Newton was the first to succeed in finding a clearly formulated basis from which he could deduce a wide field of phenomena by means of mathematical thinking—logically, quantitatively, and in harmony with experience.

On Newton's 300th birthday, *Manchester Guardian*, Christmas 1942

Newton . . . you found the only way that was available at your time to a man of the highest reasoning and creative powers. The concepts you created are even still today a part of our thinking in physics, although we know that they will have to be superseded if we are to strive for a more profound understanding of relationships.

From "Autobiographical Notes," in Schilpp, *Albert Einstein: Philosopher-Scientist*, 30–33

ON EMMY NOETHER

On receiving the new work from Fräulein Noether, I again find it a great injustice that she cannot lecture officially. I would be very much in favor of taking energetic steps in the process [to overturn this rule].

To Felix Klein, December 27, 1918, on the brilliant mathematician who was not allowed to be on the faculty of the University of Göttingen because she was a woman. Einstein Archive 14-459; *CPAE*, Vol. 8, Doc. 677

It would not have done the Old Guard at Göttingen any harm, had they learned a thing or two from her. She certainly knows what she is doing.

To David Hilbert, May 24, 1918. Einstein Archive 13-124; *CPAE*, Vol. 8, Doc. 548

In the judgment of the most competent living mathematicians, Fräulein Noether was the most significant creative mathematical genius thus far produced since the higher education of women began.

To the *New York Times* upon the death of Emmy Noether, May 4, 1935

ON MAX PLANCK

* It was largely because of the decisive and cordial manner in which he supported this theory that it attracted notice so quickly among my colleagues in the field.

Speaking about the special theory of relativity, in "Max Planck as Scientist" (1913), in *CPAE*, Vol. 4, Doc. 23. Planck was more responsible than anyone else for firmly establishing relativity theory after 1905.

How different, and how much better, it would be for mankind if there were more like him among us. But it seems not possible. Honorable persons in every age and everywhere have remained isolated, unable to influence external events.

To Frau Planck, November 10, 1947, about her husband, the German physicist and Nobel laureate for 1918. Einstein Archive 19-406

He was one of the finest people I have ever known . . . but he really did not understand physics, [because] during the eclipse of 1919 he stayed up all night to see if it would confirm the bending of light by the gravitational field. If he had *really* understood the general theory of relativity, he would have gone to bed the way I did.

> Quoted by Ernst Straus in French, *Einstein: A Centenary Volume*, 31

ON WALTHER RATHENAU

* It is easy to be an idealist when one lives in the clouds. But he was an idealist who lived down on earth and knew its scents like few others.

> On the German foreign minister who was assassinated in 1922 by members of a secret fascist terrorist organization called "Organisation Consul." Quoted in *Neue Rundschau* 33 (1922), 815–816

ON ROMAIN ROLLAND

* He is right in attacking individual greed and the national scramble for wealth that make war inevitable. He may not be far wrong in turning to social revolution as the only means of breaking the war system.

> From an interview, *Survey Graphic* 24 (August 1935), 384, 413, on the most prominent pacifist of the time

ON FRANKLIN D. ROOSEVELT

No matter when this man might have left us, we would have felt that we had suffered an irreplaceable loss. . . . May he have a lasting influence on the hearts and minds of men!

> Statement upon the president's death, in *Aufbau* (New York), April 27, 1945. According to the *New York Times*, August 19, 1946, Einstein was sure that FDR would have forbidden the bombing of Hiroshima had he been alive. Einstein had written a letter to FDR in March 1945 warning him of the bomb's devastating effects; the president died before he had a chance to read it. See Appendix.

I'm so sorry that Roosevelt is president—otherwise I would visit him more often.

> To friend Frieda Bucky. Quoted in *The Jewish Quarterly* 15, no. 4 (Winter 1967–68), 34

ON BERTRAND RUSSELL

The clarity, certainty, and impartiality you apply to the logical, philosophical, and human issues in your books are unparalleled in our generation.

> To Bertrand Russell, October 14, 1931. Einstein Archive 33-155, 75-544; also quoted in Grüning, *Ein Haus für Albert Einstein*, 369

Great spirits have always encountered opposition from mediocre minds. The mediocre mind is incapable of understanding the man who refuses to bow blindly to conventional prejudices and chooses instead to express his opinions courageously and honestly.

On controversy surrounding Russell's
appointment to the faculty of the City
University of New York. Quoted in the
New York Times, March 13, 1940

ON ALBERT SCHWEITZER

* He is a great figure who bids for the moral leadership of the world.

From an interview, *Survey Graphic* 24 (August
1935), 384, 413

He is the only Westerner who has had a moral effect on this generation comparable to Gandhi's. As in the case of Gandhi, the extent of this effect is overwhelmingly due to the example he gave by his own life's work.

Unpublished statement, 1953, originally intended
for the reedition of *Mein Weltbild*. Quoted in Sayen,
Einstein in America, 296

ON GEORGE BERNARD SHAW

* Shaw is undoubtedly one of the world's greatest figures. I once said of him that his plays remind me of Mozart. There is not one superfluous word in Shaw's prose, just as there is not one superfluous note in Mozart's music.

    In *Cosmic Religion* (1931), 109

ON UPTON SINCLAIR

* He is in the doghouse here because he relentlessly sheds light on the hurly-burly dark side of American life.

    To the Lebach family, January 16, 1931. Einstein Archive 47-373

ON SPINOZA

Spinoza is one of the most profound and pure people that our Jewish race has produced.

    In a letter of 1946. Quoted in Dürrenmatt, *Albert Einstein: Ein Vortrag*, 22

* I am fascinated with Spinoza's pantheism, but admire even more his contribution to modern thought

because he is the first philosopher to deal with the soul and body as one, and not two separate things.

Quoted in G. S. Viereck, *Glimpses of the Great* (New York, 1930)

* In my opinion, his opinions have not gained general acceptance by all those striving for clarity and logical rigor only because they require not only consistency of thought but also unusual integrity, magnanimity—and modesty.

To D. Runes, September 8, 1932. Einstein Archive 33-286

ON RABINDRANATH TAGORE

* The verbal dialogue with Tagore was a complete disaster because of difficulties in communication, and should never have been published.

To Romain Rolland, October 10, 1930, on the conversation with Indian mystic, poet, and musician Tagore that took place on July 14, 1930, in Caputh, and at which two of Tagore's secretaries took copious notes, publishing them in *Asia* 31 (1930), 138–142. Einstein shortened Tagore's first name and nicknamed him "Rabbi Tagore." Einstein Archive 33-029

ON TOLSTOY

I doubt if there has been a true moral leader of worldwide influence since Tolstoy. . . . He remains in many ways the foremost prophet of our time. . . . There is no one today with Tolstoy's deep insight and moral force.

> From an interview, *Survey Graphic* 24 (August 1935), 384, 413

ON CHAIM WEIZMANN

The chosen one of the chosen people.

> To Chaim Weizmann, October 27, 1923. Einstein Archive 33-366

* My feelings toward Weizmann are ambivalent, as Freud would say.

> Said to Abraham Pais, 1947. See Pais, *A Tale of Two Continents*, 228

ON WOODROW WILSON

* Among the most important American statesmen it is probably Wilson who most represents the intel-

lectual type. He, too, seemed not to have been very talented when it came to people skills.

From a draft manuscript, ca. 1940. Kaller's
Autographs catalog, "Jewish Visionaries," 35

On Germans and Germany

Five Nobel Prize winners in Berlin, 1931.
Left to right: Walther Nernst, Einstein, Max Planck, Robert Millikan (an American from Caltech), and Max von Laue. (Courtesy of the Archives, California Institute of Technology)

* These are not really people with natural sentiments; they are cold and of an odd mixture of snobbery and servility, without showing benevolence toward fellow humans. Ostentatious luxury alongside destitution in the streets.

> To Michele Besso, May 13, 1911, commenting on the Germans of Prague. *CPAE*, Vol. 5, Doc. 267

* Now I understand the complacency of the citizens of Berlin. One gets so much outside stimulation here that one doesn't feel one's own emptiness as profoundly as one would in a more tranquil little spot.

> To the Hurwitz family, May 4, 1914. *CPAE*, Vol. 8, Doc. 6

* It is extraordinarily inspiring here in Berlin.

> To Wilhelm Wien, June 15, 1914. *CPAE*, Vol. 8, Doc. 14

* The people are like anywhere else, but you can make a better selection because there are so many of them.

> To Robert Heller, July 20, 1914, on Berlin. *CPAE*, Vol. 8, Doc. 25

* The country is like a man with a badly upset stomach who has not yet vomited enough.

> To Aurel Stodola, March 31, 1919, on the
> right-wing putsch in Berlin two weeks earlier.
> Einstein Archive 22-257

Berlin is the place to which I am most closely bound by human and scientific ties.

> To K. Haenisch, Prussian Minister of Education,
> September 8, 1920. Einstein Archive 36-022

Germany had the misfortune of becoming poisoned, first because of plenty, and then because of want.

> Aphorism, 1923. Einstein Archive 36-591

* Funny people, these Germans. To them I am a stinking flower, yet they make me into a boutonniere time and time again.

> Travel Diary, April 17, 1925

The statements I have issued to the press were concerned with my intention to resign my position in the Academy and renounce my Prussian citizenship. I gave as my reason for these steps that I did not wish to live in a country where the individual does not enjoy equality before the law and freedom to say and teach what he likes.

To the Prussian Academy of Sciences, April 5,
1933. Einstein Archive 29-295

You have also remarked that a "good word" on my
part for "the German people" would have produced a great effect abroad. To this I must reply
that such testimony as you suggest would have
been equivalent to a repudiation of all the notions
of justice and liberty for which I have stood all of
my life. Such testimony would not be, as you put it,
a good word for the German nation.

Reply to the Prussian Academy of Sciences,
April 12, 1933, after it accepted Einstein's
resignation. Einstein Archive 29-297

* I have now been promoted to being an evil monster
in Germany, and all of my money has been taken
away. But I console myself with the thought that it
would soon have been spent, anyway.

To Max Born, May 30, 1933, after his German bank
account had been confiscated. In Born, *Born-
Einstein Letters*, 114

I cannot understand the passive response of the
whole civilized world to this modern barbarism.
Doesn't the world see that Hitler is aiming for war?

Quoted by a reporter for *Bunte Welt* (Vienna),
October 1, 1933; also quoted in Pais, *Einstein Lived
Here*, 194

The overemphasized military mentality in the German state was alien to me even as a boy. When my father moved to Italy, he took steps, at my request, to have me released from German citizenship because I wanted to become a Swiss citizen.

> 1933. Quoted in Hoffmann, *Albert Einstein: Creator and Rebel*, 26

Germany the way it used to be was [a cultural] oasis in the desert.

> To Alfred Kerr, July 1934. Einstein Archive 50-687

* Germany is still war-minded and conflict is inevitable. The nation has been on the decline mentally and morally since 1870. Many of the men I associated with in the Prussian Academy have not been of the highest caliber in the nationalistic years since the world war.

> From an interview, *Survey Graphic* 24 (August 1935), 384, 413

For centuries . . . the Germans have been trained in hard work and were made to learn many things, but they have also been trained in slavish submission, military routine, and brutality.

> From an unpublished manuscript, 1935. Quoted in Nathan and Norden, *Einstein on Peace*, 263

They have always had the tendency to treat psychopaths like knights. But they have never been able to accomplish it so successfully as at the present time.

> Written as a scribble on the reverse of a letter
> dated July 28, 1939, alluding to Hitler. Einstein
> Archive 53-160

Because of their wretched traditions, the Germans are so evil that it will be very difficult to remedy the situation by sensible, not to say humane, means. I hope that by the end of the war they will largely kill themselves off with the kindly help of God.

> To Otto Juliusburger, Summer 1942. Einstein
> Archive 38-199; also quoted in Sayen, *Einstein in
> America*, 146

The Germans as an entire people are responsible for these mass murders and must be punished as a people. . . . Behind the Nazi party stand the German people who elected Hitler after he had, in his book and in his speeches, made his shameful intentions clear beyond the possibility of misunderstanding.

> On the heroes of the Warsaw Ghetto, in *Bulletin of
> the Society of Polish Jews* (New York), 1944

Since the Germans massacred my Jewish brethren in Europe, I will have nothing further to do with

Germans, including a relatively harmless academy.
This does not include those few who remained
levelheaded within the range of possibility.

To Arnold Sommerfeld, December 14, 1946.
Einstein Archive 21-368. Einstein included Otto
Hahn, Max von Laue, Max Planck, and Arnold
Sommerfeld among the few.

The crime of the Germans is truly the most abominable ever to be recorded in the history of the so-
called civilized nations. The conduct of the German
intellectuals—seen as a group—was no better than
that of the mob.

To Otto Hahn, January 26, 1949. Einstein Archive
12-072

* The attitude of the overwhelming majority of the
Germans toward our people was such that we cannot help consider them anything but a danger. I
judge the relations of the Germans to other nations
to be equally dangerous.

From an interview with Alfred Werner, *Liberal
Judaism* 16 (April–May 1949), 4–12

Einstein with Charlie Chaplin at the premiere of the film *City Lights*, Los Angeles, January 1931. (Courtesy of Leo Baeck Institute, New York)

* At times such as this, one realizes what a sorry species one belongs to. I am moving along quietly with my contemplations while experiencing a mixture of pity and revulsion.

    To Paul Ehrenfest, August 19, 1914, at the onset of
    World War I. *CPAE*, Vol. 8, Doc. 34

* I see that often the most power-hungry and politically extreme people could not as much as kill a fly in their personal lives.

    To Paul Ehrenfest, June 3, 1917. *CPAE*, Vol. 8,
    Doc. 350

* Man tries to fashion for himself a simplified and intelligible picture of the world; he then tries to substitute this cosmos for his own world of experience. . . . Each makes this cosmos and its construction the pivot of his emotional life in order to find the peace and security he can't find in the narrow whirlpool of personal experience.

    Address to the German Physical Society on
    the occasion of Max Planck's sixtieth birthday,
    April 26, 1918. See *Mein Weltbild*, 107–110

* Failure and deprivation are the best educators and purifiers.

> To Guste Hochberger, July 30, 1919. Einstein
> Archive 43-915

Children don't heed the life experiences of their parents, and nations ignore history. Bad lessons always have to be learned anew.

> Aphorism, October 12, 1923. Einstein Archive
> 36-589

* Why do people speak of great men in terms of nationality? Great Germans, great Englishmen? Goethe always protested against being called a German poet. Great men are simply men and are not to be considered from the point of view of nationality, nor should the environment in which they were brought up be taken into account.

> *New York Times*, April 18, 1926, 12:4

It is people who make me seasick—not the sea. But I am afraid that science has yet to find a solution for this ailment.

> To Herr Schering-Kahlbaum, November 28, 1930.
> Einstein Archive 36-531

Enjoying the joys of others and suffering with them—these are the best guides for men.

To Valentine Bulgakov, November 4, 1931.
Einstein Archive 45-702

* Everything that men do or think concerns their personal needs or an escape from pain.

In *Cosmic Religion* (1931), 43

The true value of a human being is determined primarily by how he has attained liberation from the self.

Ca. June 1932, from a typescript by Jagdish Mehra
on Einstein's philosophy of life. Einstein Archive
60-492; published in *Mein Weltbild*, 10; reprinted in
Ideas and Opinions, 12

* The minority, presently the ruling class, has the school and the press, and usually the Church as well, under its thumb. This enables it to organize and sway the emotions of the masses, and use them as its tool.

To Sigmund Freud, July 30, 1932. In *Why War?* 5

* In the last analysis, everyone is a human being, whether he is an American or a German, a Jew or a Gentile. If it were possible to hold only this worthy point of view, I would be a happy man.

To Gerald Donahue, April 3, 1935. Einstein
Archive 49-502

* Man owes his strength in the struggle for existence to the fact that he is a social animal. Just like a battle between single ants on an ant hill is essential for an ant's survival, so it is also the case with individual members of a human community.

> Address for a convention at the State University of
> New York in Albany, October 15, 1936. In *School
> and Society* 44 (1936), 589–592. See also the *New York
> Times*, October 16, 1936, 11:1; published as
> "On Education," in *Out of My Later Years*, 36

* When one looks at humankind today, one notices with regret that quantity does not make up for quality: if quantity could only substitute for quality, we would be in better circumstances now than was Ancient Greece.

> To Ruth Norden, December 21, 1937. Kaller's
> Autographs catalog, "Jewish Visionaries," 33

Common convictions and aims, similar interests, will in every society produce groups that, in a certain sense, act as units. There will always be friction between such groups—the same sort of aversion and rivalry that exists between individuals.... In my opinion, uniformity in a population would not be desirable, even if it were attainable.

> From "Why Do They Hate the Jews?" *Collier's*
> magazine, November 26, 1938

It is better for people to be like the beasts ... they should be more intuitive; they should not be too conscious of what they are doing while they are doing it.

From a conversation recorded by Algernon Black,
Fall 1940. Einstein Archive 54-834

We have to do the best we can. This is our sacred human responsibility.

Ibid.

* Perfection of means and confusion of goals seem, in my opinion, to characterize our age.

Broadcast recording for a science conference in
London, 1940 or 1941

* Perhaps there is something benevolent in the wasteful sport that Nature, seemingly blindly, places on her creatures. It can be beneficial only if we persuade young people how critical that decision [marriage and reproduction] is—often made at the moment when Nature leaves us in a kind of drunken sensual delusion so that we least have our good judgment when we most need it.

To Hans Muehsam, June 4, 1946, on his rejection of
any concerted effort to "improve" the human race.
Einstein Archive 38-356

* It is easier to denature plutonium than to denature the evil spirit of man.

> From an interview, *New York Times*, June 23, 1946.
> Also in Nathan and Norden, *Einstein on Peace*, 385

There is only one road to human greatness: through the school of hard knocks.

> Comment on W. White's article, "Why I Remain a Negro," October 1947. Einstein Archive 59-009

* Man is, at one and the same time, a solitary being and a social being. As a solitary being, he attempts to protect his own existence and the existence of those who are closest to him, to satisfy his personal desires, and to develop his innate abilities. As a social being, he seeks to gain the recognition and affection of his fellow human beings, to share in their pleasures, to comfort them in their sorrows, and to improve their conditions in life.

> From "Why Socialism?" *Monthly Review*, May 1949. (Thanks to Eric Seele for this contribution.)

* You must be aware that most men (and also not only a few women) are by nature not monogamous. This nature makes itself even more forceful when tradition and circumstance stand in an individual's way.

To Eugenie Anderman, an unknown woman who
was seeking advice from Einstein regarding her
husband's infidelity, June 2, 1953. (Courtesy of
Andor Carius.)

* A forced faithfulness is a bitter fruit for all concerned.

Ibid.

We all are nourished and housed by the work of
our fellowmen and we have to pay honestly for it
not only by the work we have chosen for the sake of
our inner satisfaction but by work which, according
to general opinion, serves them. Otherwise we become parasites, however modest our needs may be.

To a man who wanted to spend his time being
subsidized to study rather than work, July 28, 1953.
Einstein Archive 59-180

To obtain an assured favorable response from people, it is better to offer them something for their
stomachs instead of their brains.

To a chocolate manufacturer, March 19, 1954.
Einstein Archive 60-401

Fear or stupidity has always been the basis of most
human actions.

To E. Mulder, April 1954. Einstein Archive 60-609

Einstein with (*left to right*) Zionists Benzion Mossensom, Chaim Weizmann (the future president of Israel), and Menachem Ussishkin, on arrival in the United States, April 2, 1921, aboard the SS *Rotterdam*. (Copyright © Bettmann/Corbis)

In his youth, Einstein did not identify strongly with Jewish culture and religion. His parents were assimilated Jews in southern Germany who had distanced themselves from their Jewish roots and were more interested in the realities of entrepreneurship and making a good living. However, he did receive private religious instruction at home and at first embraced it intensely, only to reject it decisively by the age of twelve as his interest in science became stronger. He then declared himself "without religious affiliation." He "rediscovered" Judaism again in the 1920s following his retreat from intense scientific investigations. This was coincident both with the advent of anti-Semitism and Zionism and with the confirmation of his general theory of relativity, which brought him worldwide fame. Einstein's Zionism was cultural rather than political. It emphasized the cultural and spiritual renewal of the Jewish people, as opposed to political Zionism, which focused on the establishment of a Jewish state. Still, he supported the creation of Israel as a refuge for Jews because he believed in the power of a community as a cohesive force. See Schulmann, "Einstein Rediscovers Judaism," and Stachel, "Einstein's Jewish Identity."

* It goes against the grain to travel without necessity to a country in which my kinsmen have been persecuted so brutally.

To P. P. Lazarev, May 16, 1914, responding to an invitation from the Imperial Academy of Sciences

in St. Petersburg to come to Russia to watch the
1914 lunar eclipse. Quoted in Winter and Jarosch,
eds., *Wegbereiter der deutsch-slawischen
Wechselseitigkeit* (Berlin, 1983), 354

* I am neither a German citizen, nor is there anything
in me that can be described as a "Jewish faith."
However, I am happy to be a member of the Jewish
people, even though I do not regard them as the
Chosen People.

To Central-Verein Deutscher Staatsbürger
Jüdischen Glaubens, April 5, 1920. Published in
Israelitisches Wochenblatt, September 24, 1920.
CPAE, Vol. 7, forthcoming

* Perhaps it is thanks to anti-Semitism that we can
preserve ourselves as a race: at least, this is what I
believe.

Ibid.

I am not at all eager to go to America but am doing
it only in the interests of the Zionists, who must beg
for dollars to build educational institutions in
Jerusalem and for whom I act as high priest and
decoy. . . . But I do what I can to help those in my
tribe who are treated so badly everywhere.

To Maurice Solovine, March 8, 1921. Published in
Letters to Solovine, 41

* Despite my internationalist beliefs I have always felt an obligation to stand up for the persecuted and morally oppressed members of my tribe as much as I am able to.

> To Fritz Haber, March 9, 1921. Einstein Archive 36-840

* I know of no public event that has given me such pleasure as the proposal to establish a Hebrew University in Jerusalem. The traditional respect for knowledge that Jews have maintained intact through many centuries of severe hardship has made it particularly painful for us to see so many talented sons of the Jewish people cut off from higher education.

> From an interview, *New York Times*, April 3, 1921. See similar remarks with respect to Brandeis University, below.

* Zionism really represents a new Jewish ideal, one that can give the Jewish people renewed joy in existence. . . . I am pleased that I accepted Weizmann's invitation.

> To Paul Ehrenfest, June 18, 1921, on his fundraising tour of the United States with Chaim Weizmann on behalf of the Hebrew University. Einstein Archive 9-561

* Anti-Semitism in Germany also has effects that, from a Jewish point of view, should be welcomed.

I believe German Jewry owes its continued existence to anti-Semitism. . . . Without this distinction,
the assimilation of Jews in Germany would happen
quickly and unimpeded.

> *Jüdische Rundschau*, June 21, 1921. See also *CPAE*,
> Vol. 7, forthcoming

* So long as I lived in Switzerland, I did not become aware of my Jewishness. . . . This changed as
soon as I took up residency in Berlin. . . . I saw how
anti-Semitism prevented Jews from pursuing or-
derly studies, and how they struggled to secure a
livelihood.

> Ibid.

* I am not a Jew in the sense that I would demand the
preservation of the Jewish or any other nationality
as an end in itself. Rather, I see Jewish nationality
as a fact and I believe that every Jew must reap the
consequences of this fact.

> Ibid.

* My Zionism does not exclude internationalism.

> Ibid.

* I consider the raising of Jewish self-confidence as a
necessary fact, also in the interest of living normally

together with non-Jews. This was the major motive of my joining the Zionist movement. . . . Zionism strengthens the self-confidence of Jews, which is necessary for their existence in the diaspora, and a Jewish center in Palestine creates a strong bond that gives Jews moral support.

Ibid.

* Zionism, to me, is not just a colonizing movement to Palestine. The Jewish nation is a living fact in Palestine as well as in the diaspora, and Jewish feelings must be kept alive wherever Jews live. . . . I believe that every Jew has duties toward his fellow Jews.

Ibid.

* By repatriating Jews to Palestine and giving them a healthy and normal economic existence, Zionism is a productive activity that enriches human society.

Ibid.

Where dull-witted clansmen of our tribe were praying aloud, their faces turned to the wall, their bodies swaying to and fro. A pathetic sight of men with a past but without a future.

On his visit to the Wailing Wall in Jerusalem, February 3, 1923, recorded in his travel diary. Einstein Archive 29-129 to 29-131

The heart says yes, but the mind says no.

> Travel Diary, February 13, 1923, on the invitation
> to accept a position at the Hebrew University of
> Jerusalem. In ibid. Also quoted in Hoffmann,
> "Einstein and Zionism," 241

* Our Jews are up to much down there [Jerusalem]
and are fighting among one another as usual. I
actually have quite a bit to do with all of this
because—as you know—I have become a Jewish
saint.

> To Michele Besso, December 25, 1925. Einstein
> Archive 7-356. The tongue-in-cheek "Jewish saint"
> reference also appears in a letter to sons
> Hans Albert and Eduard, November 24, 1923 (see
> above, "On Einstein Himself"). In a letter to Paul
> Ehrenfest of May 4, 1920, he joked that "the bones
> of the saint were requested to be present" at a
> conference in Halle later that month.

* A Jew who strives to impregnate his spirit with humanitarian ideals can declare himself a Zionist
without contradiction. What one must be thankful
for to Zionism is that it is the only movement that
has given Jews a justified pride.

> In *Jüdische Rundschau* 30 (1925), 129

* If we did not have to live among intolerant, nar-
row-minded, and violent people, I would be the
first to discard all nationalism in favor of a universal humanity.

To Professor Hellpack, October 8, 1929; Einstein
Archive 46-657

Should we be unable to find a way to honest cooperation and honest pacts with the Arabs, then we
shall have learned nothing from our 2,000 years of
suffering and will deserve our fate.

To Chaim Weizmann, November 25, 1929. Einstein
Archive 33-411

Jewry has proved throughout history that the intellect is the best weapon. . . . It is our duty as Jews
to put at the disposal of the world our several-
thousand-year-old sorrowful experience and, true
to the ethical traditions of our forefathers, become
soldiers in the fight for peace, united with the noblest elements in all cultural and religious circles.

From an address at a Jewish meeting in Berlin,
1929. Quoted in Frank, *Einstein: His Life and Times*,
156

The Jewish religion is . . . a way of sublimating
everyday existence. . . . It demands no act of faith—
in the popular sense of the term—on the part of its
members. And for that reason there has never been
a conflict between our religious outlook and the
world outlook of science.

In *Forum and Century* 83 (1930), 373

I have always been annoyed by the undignified as-
similationist cravings and strivings that I have observed in so many of my [Jewish] friends. . . . These
and similar happenings have awakened in me the
Jewish national sentiment.

> In *About Zionism* (1931)

* I did not discover the Jewish people until I was in
America. I had *seen* many Jews, but neither in Berlin nor elsewhere in Germany had I ever encountered the Jewish people.

> Ibid., 48

Judaism is not a creed: the Jewish God is simply a
negation of superstition, an imaginary result of its
elimination. It is also an attempt to base moral law
on fear, a regrettable and discreditable attempt. Yet
it seems to me that the strong moral tradition of the
Jewish nation has to a large extent shaken itself free
from this fear. It is clear also that "serving God"
was equated with "serving the living." The best of
the Jewish people, especially the Prophets and
Jesus, contended tirelessly for this.

> From "Is There a Jewish Point of View?" *Opinion*,
> September 26, 1932; reprinted in *Ideas and Opinions*,
> 186

* Judaism seems to me to be concerned almost exclusively with the moral attitude in life and toward

life. . . . The essence of that conception seems to me
to lie in an affirmative attitude toward the life of all
creation.

> Ibid., 185. As reflected in this and similar
> statements, Einstein rejected any racial or other
> biological sanctions for Judaism; rather, he saw it
> primarily as an attitude toward life. See Stachel,
> "Einstein's Jewish Identity," for a discussion about
> Einstein's rejection of any taint of race in his
> concept of Judaism.

* The Jews resemble an uncondensable noble gas that
can assume a substantial form of existence only by
adhering to a firm object. This applies to me as well.
But perhaps it is in this very chemical inertia that
our ability to act and to persist lies.

> To Paul Ehrenfest, June 1933. Einstein Archive
> 10-260

The pursuit of knowledge for its own sake, an almost fanatical love of justice, and the desire for personal independence—these are the features of the
Jewish tradition which make me thank my lucky
stars that I belong to it.

> From "Jewish Ideals," in *Mein Weltbild* (1934), 89;
> reprinted in *Ideas and Opinions*, 185

* It is of no use to try to convince others of our spiritual and intellectual equality by way of reason,
since the others' attitude does not come by way of

their brains. We must instead emancipate ourselves socially and satisfy our own social requirements.

From "Anti-Semitism and Academic Youth," in
ibid., 188

There are no German Jews, there are no Russian Jews, there are no American Jews. Their only difference is their daily language. There are in fact only Jews.

From a speech at a Purim dinner at the German-
Jewish Club in New York, March 24, 1935,
reprinted shortly thereafter in the *New York Herald
Tribune*. A reader took issue with this statement
and asked Einstein to explain what he meant. See
the next quotation for a partial answer.

* Seen through the eyes of the historian, their history of suffering teaches us that the fact of being a Jew has had a greater impact than the fact of belonging to political communities. If, for example, the German Jews were driven from Germany, they would cease to be Germans and would change their language and their political affiliation; but they would remain Jews. . . . I see the reason not so much in racial characteristics as in firmly rooted traditions that are by no means limited to the area of religion.

To Gerald Donahue, April 3, 1935. Einstein
Archive 49-502. Note, however, that Einstein

himself continued to speak the German language
even after he was driven from Germany.

The intellectual decline brought on by shallow ma-
terialism is a far greater menace to the survival of
the Jew than the numerous external foes who
threaten his existence with violence.

Quoted in the *New York Times*, June 8, 1936

I should much rather see a reasonable agreement
with the Arabs based on living together in peace
than the creation of a Jewish state.

From a speech entitled "Our Debt to Zionism,"
before the National Labor Committee for Palestine
on April 17, 1938, in New York. Full text published
in *New Palestine* (Washington, D.C.), April 28, 1938

Judaism owes a great debt of gratitude to Zionism.
The Zionist movement has revived among Jews a
sense of community. It has performed productive
work . . . in Palestine, to which self-sacrificing Jews
throughout the world have contributed. . . . In particular, it has been possible to lead a not inconsider-
able part of our youth toward a life of joyous and
creative work.

Ibid.

My awareness of the essential nature of Judaism re-
sists the idea of a Jewish state with borders, an
army, and a measure of temporal power.... I am
afraid of the inner damage Judaism will sustain—
especially from the development of a narrow nationalism within our ranks, which we have already
had to fight strongly even without a Jewish state.
... A return to a nation in the political sense of the
word would be equivalent to turning away from
the spiritualization of our community that we owe
to the genius of our prophets.

> Ibid.

The Jews as a group may be powerless, but the sum
of the achievements of their individual members is
everywhere considerable and telling, even though
those achievements were made in the face of
obstacles.

> From "Why Do They Hate the Jews?" *Collier's*
> magazine, November 26, 1938

[The Nazis] see the Jews as a nonassimilable element that cannot be driven into uncritical acceptance and that ... threatens their authority because
of its insistence on popular enlightenment of the
masses.

> Ibid.

The Jew who abandons his faith (in the formal sense of the word) is in a position similar to a snail that abandons its shell. He remains a Jew.

Ibid.

The bond that has united the Jews for thousands of years and that unites them today is, above all, the democratic ideal of social justice, coupled with the ideal of mutual aid and tolerance among all men. . . . The second characteristic of Jewish tradition is the high regard in which it holds every form of intellectual aspiration and spiritual effort.

Ibid.

Zionism is nationalism whose aim is not power but dignity.

In an article in the *New York Times Magazine*,
March 12, 1944. Einstein Archive 29-102

Zionism gave the German Jews no great protection against annihilation. But it did give the survivors the inner strength to endure the debacle with dignity and without losing their healthy self-respect.

To an anti-Zionist Jew, probably January 1946.
Quoted in Dukas and Hoffmann, *Albert Einstein,
the Human Side*, 64

* Under existing conditions, our young scientific talents frequently have no access to scholarly professions, which means that our proudest tradition—the appreciation of productive work—would be faced with slow extinction if we remain as inactive in this area as in the past.

> To David Lilienthal, July 1946, on his approval of the founding of Brandeis University to serve Jewish students. Within a year he added that it should "not know of discrimination for or against anybody because of sex, color, creed, national origin, or political opinion." In 1953 he refused an honorary doctorate from the institution because of some personal disputes with the founders in 1947. Einstein Archive 40-398, 40-432. See also S. S. Schweber, "Albert Einstein and the Founding of Brandeis University," unpublished manuscript

* The plight of the surviving victims of German persecution bears witness to the degree to which the moral conscience of mankind has weakened. Today's meeting shows that not all men are prepared to accept the horror in silence.

> Message on the dedication of the Riverside Drive Memorial, New York, to the victims of the Holocaust, October 19, 1947. Rosenkranz, *Albert through the Looking-Glass*, 93

The wisdom and moderation the leaders of the new state have shown gives me confidence that gradually relations will be established with the Arab people which are based on fruitful cooperation and mutual respect and trust. For this is the only means through which both peoples can attain true independence from the outside world.

> Statement to the Hebrew University of Jerusalem upon receipt of an honorary doctorate, March 15, 1949. Einstein Archive 28-854, 37-296

This university is today a living thing, a home of free learning and teaching and happy collegial work. There it is, on the soil that our people have liberated under great hardships; there it is, a spiritual center of a flourishing and buoyant community whose accomplishments have finally met with the universal recognition they deserve.

> Ibid.

* Zionism in 1921 strove for the establishment of a national home, not the foundation of a state in the political sense. However, this latter aim has been realized because of the pressure of necessity rather than emergency. To discuss this development retrospectively seems to be academic. As far as the attitude commonly described as "orthodoxy" is concerned, I never had much sympathy for it. Nor do I think it plays an important role now, or that it is likely to in the future.

> In answer to interviewer Alfred Werner's question about his support of Zionism in the establishment

of Israel as a secular state, in *Liberal Judaism* 16
(April–May 1949), 4–12

* I must be very careful not to do any foolish thing or
to write any foolish book in order to live up to that
distinction. I am proud of the honor, not on my
own account, but because I am a Jew. It certainly
denotes progress when a Christian church honors a
Jewish scientist.

> Referring to his sculptured image over the main
> entrance of Riverside Church in New York City
> (see photo in "Religion" section), which also
> depicts other immortal leaders of humanity. Ibid.
> Riverside Church, modeled after the thirteenth-
> century cathedral in Chartres, France, was
> completed in 1930 and is interdenominational.

The Jews of Palestine did not fight for political independence for its own sake, but they fought to
achieve free immigration for the Jews of many
countries where their very existence was in danger;
free immigration also for all those who were longing for a life among their own. It is no exaggeration
to say that they fought to make possible a sacrifice
that is perhaps unique in history.

> From an NBC radio broadcast for the United
> Jewish Appeal Conference, Atlantic City,
> November 27, 1949. Einstein Archive 58-904

The support for cultural life is of primary concern to the Jewish people. We would not be in existence today as a people without this continued activity in learning.

> Statement on the occasion of the twenty-fifth anniversary of the Hebrew University of Jerusalem. Quoted in the *New York Times*, May 11, 1950

* Jewry is a group of people with a common history and with certain traditions besides the religious one. They are united by common interests created and maintained by the outside world through a mostly antagonistic attitude called prejudice.

> To Alan E. Mayers, a Princeton University student, October 20, 1950. (Thanks to Mr. Mayers for sending the letter to me.)

My relationship to the Jewish people has become my strongest human bond, ever since I became fully aware of our precarious situation among the nations of the world.

> Statement to Abba Eban, Israeli ambassador to the United States, November 18, 1952. Einstein Archive 28-943

For the young state to achieve real independence, and conserve it, a group of intellectuals and experts must be produced in the country itself.

> Quoted in the *New York Times*, May 25, 1953

Israel is the only place on earth where Jews have the possibility to shape public life according to their own traditional ideals.

From an address at a planning conference of American Friends of the Hebrew University, in Princeton, N.J., September 19, 1954. Einstein Archive 28-1054

*The most important aspect of our [Israel's] policy must be our ever-present, manifest desire to institute complete equality for the Arab citizens living in our midst. . . . The attitude we adopt toward the Arab minority will provide the real test of our moral standards as a people.

To Zvi Lurie, January 5, 1955, written three months before Einstein's death

Only when we [Jews] have the courage to regard ourselves as a nation, only when we have respect for ourselves, can we win the respect of others.

Quoted in Hoffmann, "Einstein and Zionism," 237

Zionism indeed represents a new Jewish ideal that we can restore to the Jewish people their joy in existence.

Quoted in Dukas and Hoffmann, *Albert Einstein, the Human Side*, 63

If I were to be president, sometimes I would have to say to the Israeli people things they would not like to hear.

To Margot Einstein, on his decision to turn down the presidency of Israel. Quoted in Sayen, *Einstein in America*, 247

Einstein in middle age. (Library of Congress. Courtesy
of AIP Emilio Segrè Visual Archives)

If there is no price to be paid, it is also not of value.

Aphorism, June 20, 1927. Einstein Archive 35-582

I believe that a simple and unassuming life is good for everybody, both physically and mentally.

From "What I Believe," *Forum and Century* 84 (1930), 193–194; reprinted in *Ideas and Opinions*, 8–11

Only a life lived for others is a life worthwhile.

In answer to a question asked by the editors of *Youth*, a journal of Young Israel of Williamsburg, N.Y. Quoted in the *New York Times*, June 20, 1932; Einstein Archive 60-492

Strange is our situation here on earth. Each of us comes for a short visit, not knowing why, yet sometimes seeming to divine a purpose.

From "My Credo," for the German League of Human Rights, 1932; also quoted in Leach, *Living Philosophies*, 3

The life of the individual has meaning only insofar as it aids in making the life of every living thing nobler and more beautiful. Life is sacred, that is to say, it is the supreme value, to which all other values are subordinate.

From "Is There a Jewish Point of View?" Published in *Mein Weltbild* (1934), 89–90; reprinted in *Ideas and Opinions*, 185–187

When the expected course of everyday life is interrupted, we realize that we are like shipwrecked people trying to keep their balance on a miserable plank in the open sea, having forgotten where they came from and not knowing whither they are drifting.

To a couple who unexpectedly lost a child or grandchild, April 26, 1945. Einstein Archive 56-852

We must recognize what in our accepted tradition is damaging to our fate and dignity—and shape our lives accordingly.

On the American attitude toward blacks, from an address at Lincoln University upon receipt of an honorary doctorate. *New York Times*, May 4, 1946, 6; also in "The Negro Question," in *Out of My Later Years*, 128

A life directed chiefly toward the fulfillment of personal desires will sooner or later always lead to bitter disappointment.

To L. Lee, January 16, 1954. Einstein Archive 60-235

Every reminiscence is colored by the way things are today, and therefore by a delusive point of view.

From "Autobiographical Notes," in Schilpp, *Albert Einstein: Philosopher-Scientist*, 3

If you want to live a happy life, tie it to a goal, not to people or objects.

Quoted by Ernst Straus in French, *Einstein: A Centenary Volume*, 32

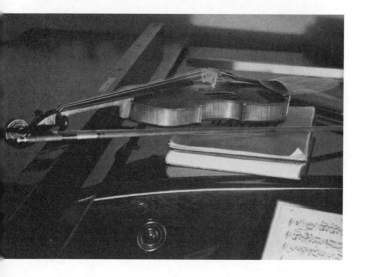

Einstein's violin, which he nicknamed "Lina."
(Courtesy of Lotte Jacobi Archives, University of
New Hampshire)

Bach, Mozart, and some old Italian and English composers were Einstein's favorites; also Schubert, because of the composer's ability to express emotion. He was considerably less fond of Beethoven, regarding his music as too dramatic and personal. Handel, he felt, was technically good but displayed shallowness. Schumann's shorter works were appealing because they were original and rich in feeling. Mendelssohn demonstrated considerable talent but lacked depth. Einstein liked some *lieder* and chamber music by Brahms. He found Wagner's musical personality indescribably offensive "so that for the most part I can listen to him only with disgust." He considered Richard Strauss gifted but without inner truth and concerned too much with outside effect. (From a response to a questionnaire, 1939. Einstein Archive 34-322)

Einstein began to play the violin at age six; by 1950 he had given it up and instead played around on the piano. He had called his violin "Lina" and willed it to his grandson, Bernhard. See Frank, *Einstein: His Life and Times*, 14; Grüning, *Ein Haus für Albert Einstein*, 251

* Stick to the Mozart sonatas. Your papa, too, learned to know music well through them.

To son Hans Albert, January 8, 1917. Einstein had "fallen in love with the Mozart sonatas" at age thirteen. *CPAE*, Vol. 8, Doc. 287, n. 2

* The differences between Japanese music and our own are truly fundamental. Whereas in our European music, chords and architectural arrangement are essential and are a given, they are absent in Japanese music. Both, however, have the same thirteen notes that make up an octave. To me, Japanese music is a painting of emotions that has a surprising and immediate effect. . . . I have the impression that it is all about giving a stylized presentation of the emotions found in the human voice, as well as the sounds of nature that stir the human soul, such as birdsong and the rumble of the ocean. This sensation is given force because percussion instruments, which are not limited by pitch and are especially well suited for rhythmic characterization, play a large role. . . . To my mind, the greatest obstacle to the acceptance of Japanese music as a great art form is its lack of formal arrangement and architectural structure.

> *Kaizo* 5, no. 1 (January 1923), 339

* Unfortunately I don't feel I am in a position, on the strength of either my sexual or my musical abilities, to accept your kind invitation.

> To K. Singer, August 16, 1926, in declining an
> invitation to participate in a musical event at the
> First International Congress for Sexual Research.
> Einstein Archive 44-905

Music does not *influence* research work, but both are nourished by the same sort of longing, and they complement each other in the release they offer.

> To Paul Plaut, October 23, 1928. Einstein Archive 28-065; also quoted in Dukas and Hoffmann, *Albert Einstein, the Human Side*, 78

* If I were not a physicist, I would probably be a musician. I often think in music. I live my daydreams in music. I see my life in terms of music. . . . I get most joy in life out of music.

> From an interview, *Saturday Evening Post*, October 26, 1929

* In Europe, music has come too far away from popular art and popular feeling and has become something like a secret art with conventions and traditions of its own.

> In conversation with Indian mystic, poet, and musician Rabindranath Tagore, discussing the possibility of self-expression in Eastern and Western music. In *Asia* 31 (March 1931)

* It requires a very high standard of art to realize fully a great idea in original music so that one can make variations upon it. [In the West], the variations are often prescribed.

> Ibid.

* The difficulty is that really good music, whether of the East or of the West, cannot be analyzed.

> Ibid.

* Even if one loves to play
  One's little fiddle night and day
  It's not right to broadcast it
  Lest the list'ners scoff at it.
  If you scratch with all your might—
  Which is certainly your right—
  Then bring down the windowpane
  So the neighbors don't complain.

> Poem to Emil Hilb, April 18, 1939 (my translation)

* I feel uncomfortable listening to Beethoven. I think he is too personal, almost naked. Give me Bach, rather, and then more Bach.

> From an interview with Lili Foldes, *The Etude*,
> January 1947

* I am done fiddling. With the passage of years, it has become more and more unbearable for me to listen to my own playing.

> To Queen Elizabeth of Belgium, January 6, 1951.
> Quoted in *Einstein on Peace*, 554

Mozart wrote such nonsense here!

> While struggling to play a piece by Mozart.
> Quoted by Margot Einstein in an interview with
> J. Sayen for *Einstein in America*, 139

First I improvise, and if that doesn't help, I seek solace in Mozart. But when I am improvising and it appears that something may come of it, I require the clear constructions of Bach in order to follow through.

> Explaining how he relaxes after work playing his
> violin in his Berlin kitchen, a room with superior
> acoustics. Quoted in Ehlers, *Liebes Hertz!* 132

* Mozart's music is so pure and beautiful that I see it as a reflection of the inner beauty of the universe.

> Quoted in Hermann, *Einstein*, 158

Einstein in old age. (Courtesy of AIP Emilio Segrè
Visual Archives, Lande Collection)

Einstein was a pacifist from his youth until 1933, when Hitler forced his hand on the issue. From 1933 to 1945, he saw some need for military action under certain circumstances; in particular, he felt that military strength among "the nations that have stayed normal" was vital against the German aggressor (Fölsing, *Albert Einstein*, 676). In general, however, he believed that a "supranational" world government was necessary to preserve civilization and individual freedoms. From 1945 until his death in 1955, he spoke out in favor of world government as a moral imperative and in support of the control of nuclear weapons.

He who cherishes the values of culture cannot fail to be a pacifist.

From a handbook on pacifism, *Die Friedensbewegung*, ed. Kurt Lenz and Walter Fabian (1922). Quoted in Nathan and Norden, *Einstein on Peace*, 55; also quoted as "A human being who considers spiritual values as supreme must be a pacifist," in Clark, *Einstein*, 48

* The field of history has by no means advanced the ideals of pacifism. Many of its representatives . . . have publicly made astounding and strongly chauvinistic and military pronouncements. . . . The situation is quite different in the natural sciences.

From *Die Friedensbewegung*, ibid.

* Because of the universal character of their subject matter and their need for internationally organized cooperation, [scientists] are inclined toward international understanding and therefore favor pacifist goals.

    Ibid.

* Technology resulting from the sciences has internationally chained together economies and this has caused all wars to become a matter of international importance. When this situation has entered the consciousness of mankind, after sufficient turmoil, then men will also find the energy and goodwill to create organizations that have the power to end wars.

    Ibid.

* No person has the right to call himself a Christian or Jew so long as he is prepared to engage in systematic murder at the command of an authority, or allow himself to be used in any way in the service of war or the preparation for it.

    Ibid.

* I would absolutely refuse any direct or indirect war service and would try to persuade my friends to do

the same, regardless of the reasons for the cause of a war.

Ibid.

In all cases where a reasonable solution of difficulties is possible, I favor honest cooperation and, if this is not possible under prevailing circumstances, Gandhi's method of peaceful resistance to evil.

Ibid. Also quoted in Nathan and Norden, *Einstein on Peace*, 596

The conscientious objector is a revolutionary. In deciding to disobey the law he sacrifices his personal interests to the most important cause of working for the betterment of society.

Ibid. Also quoted in Nathan and Norden, *Einstein on Peace*, 542–543

My pacifism is an instinctive feeling, a feeling that possesses me because the murder of people is disgusting. My attitude is not derived from any intellectual theory but is based on my deepest antipathy to every kind of cruelty and hatred.

To Paul Hutchinson, editor of *Christian Century*, July 1929. Quoted in Nathan and Norden, *Einstein on Peace*, 98; also quoted in Clark, *Einstein*, 427

I have made no secret, either privately or publicly, of any sense of outrage over officially enforced military and war service. I regard it as a duty of conscience to fight against such barbarous enslavement of the individual with every means available.

> Statement to the Danish newspaper *Politiken*,
> August 5, 1930

That a man can take pleasure in marching in formation to the strains of a band is enough to make me despise him.

> From "What I Believe," *Forum and Century* 84
> (1930), 193–194; reprinted in *Ideas and Opinions*,
> 8–11

I believe serious progress [in the abolition of war] can be achieved only when men become organized on an international scale and refuse, as a body, to enter military or war service.

> Statement in *Jugendtribüne*, April 17, 1931

There are two ways of resisting war: the legal way and the revolutionary way. The legal way involves the offer of alternative service not as a privilege for a few but as a right for all. The revolutionary view involves an uncompromising resistance, with a view to breaking the power of militarism in time of peace or the resources of the state in time of war.

> Statement in *The New World*, July 1931

I appeal to all men and women, whether they be eminent or humble, to declare that they will refuse to give any further assistance to war or the preparation of war.

In a statement to the War Resisters International, Lyons, France, 1931. Quoted in Frank, *Einstein: His Life and Times*, 158; also quoted in the *New York Times*, August 2, 1931

I believe the most important mission of the state is to protect the individual and make it possible for him to develop into a creative personality. . . . The state violates this principle when it compels us to do military service.

From *The Nation* 33 (1931), 300; also quoted in the *New York Times*, November 22, 1931

I am not only a pacifist, but a militant pacifist. I am willing to fight for peace. . . . Is it not better for a man to die for a cause in which he believes, such as peace, than to suffer for a cause in which he does not believe, such as war?

From an interview during a visit to the United States, 1931. Quoted in Alfred Lief, ed., *The Fight against War* (New York: John Day, 1933)

* We all know that when a war comes, every man accepts the duty to commit a crime—the crime of murder—each man for his own country. Those who realize the immorality of war should do their utmost to disentangle themselves from this old idea of military duty—and so become liberated from slavery.

"Militant Pacifism," in *Cosmic Religion* (1931), 58

* Peace cannot be kept by force. It can only be achieved by understanding. You cannot subjugate a nation forcibly unless you wipe out every man, woman, and child. Unless you wish to use such drastic measures, you must find a way of settling your disputes without resort to arms.

Ibid., 67

* I am the same ardent pacifist I was before. But I believe that the tool of refusing military service can be advocated again in Europe only when the military threat from aggressive dictatorships toward democratic countries has ceased to exist.

To a rabbi in Rochester, N.Y., April 5, 1934. In
Nathan and Norden, *Einstein on Peace*, 250

* Pacifism defeats itself under certain conditions, as it would in Germany today. . . . We must work with the people to create a public sentiment that will outlaw war: (1) create the idea of supersovereignty; . . . (2) face the economic causes of war.

From an interview, *Survey Graphic* 24 (August
1935), 384, 413

* In the twenties, when no dictatorships existed, I advocated that refusing to go to war would make war improper. But as soon as coercive conditions appeared in certain nations, I felt that it would weaken the less aggressive nations vis-à-vis the more aggressive ones.

New York Times, December 30, 1941

It is my belief that the problem of bringing peace to the world on a supranational basis will be solved only by employing Gandhi's method on a larger scale.

To G. Nellhaus, March 20, 1951. Einstein Archive 60-683; also quoted in Nathan and Norden, *Einstein on Peace*, 543

I can identify my views almost completely with those of Gandhi. But I would (individually and collectively) resist with violence any attempt to kill or to take away from my people or me the basic means of subsistence.

To A. Morrisett, March 21, 1952. Einstein Archive 60-595

The goal of pacifism is possible only through a supranational organization. To stand unconditionally for this cause is . . . the criterion of true pacifism.

Ibid.

The more a country makes military weapons, the more insecure it becomes: if you have weapons, you become a target for attack.

From a conversation with M. Aram, January 1953.
Einstein Archive 59-109

* I have always been a pacifist, i.e., I have declined to recognize brute force as a means for the solution of international conflicts. Nevertheless, it is, in my opinion, not reasonable to cling to that principle unconditionally. An exception has necessarily to be made if a hostile power threatens wholesale destruction of one's own group.

To H. Herbert Fox, May 18, 1954. Einstein Archive
59-727

I am a *dedicated* but not an *absolute* pacifist; this means that I am opposed to the use of force under any circumstances except when confronted by an enemy who pursues the destruction of life as an *end in itself.*

To a Japanese correspondent, June 23, 1954.
Einstein Archive 61-297

On Peace, War, the Bomb, and the Military

Einstein in 1931. (Photo by Johan Hagemeyer.
Courtesy of the Archives, California Institute of
Technology)

* Even the scholars in various lands have been acting as if their brains had been amputated.

> To Romain Rolland, March 22, 1915, on the outbreak of World War I. Rolland was the most prominent pacifist of his time. *CPAE*, Vol. 8, Doc. 65

* The psychological roots of war are, in my opinion, biologically founded in the aggressive characteristics of the male creature.

> From "My Opinion on the War," for the Goethebund of Berlin, October–November 1915. Einstein repeated this opinion in an interview with anthropologist Ashley Montagu thirty-one years later, in June 1946, claiming that a child's naughtiness and a parent's spanking—"domestic violence"—were innately reactive, instinctive acts and a microcosmic example of international violence and aggression. He was essentially agreeing with Sigmund Freud's conclusions, but Montagu disagreed, and Einstein was finally persuaded that the doctrine of man's innate depravity was unsound. See Montagu's article, "Conversations with Einstein," in *Science Digest*, July 1985

That worst outcrop of herd life, the military system, which I abhor ... this plague-spot of civilization, ought to be abolished with all possible speed. Heroism on command, senseless violence, and all the loathsome nonsense that goes on in the name of patriotism—how passionately I hate them! How vile and despicable war seems to me! I would rather be hacked into pieces than take part in such an abominable business.

From "What I Believe," *Forum and Century* 84 (1930), 193–194; reprinted in *Ideas and Opinions*, 8–11

War is not a parlor game in which the players obediently stick to the rules. Where life and death are at stake, rules and obligations go by the board. Only the absolute repudiation of all war can be of any use here.

From an address to a group of German pacifist students, ca. 1930. Published in *Mein Weltbild*, 49; reprinted in *Ideas and Opinions*, 94

* Compulsory military service, as a hotbed of unhealthy nationalism, must be abolished; most important, conscientious objectors must be protected on an international basis.

From "The Disarmament Conference of 1932," *The Nation* 133 (1931), 100; reprinted in *Ideas and Opinions*, 98

We must ... dedicate our lives to drying up the source of war: ammunition factories.

From an interview, May 23, 1932. Published in *Pictorial Review*, February 1933; quoted in Clark, *Einstein*, 453

This is the problem: Is there any way of delivering mankind from the menace of war? It is common knowledge that with the advance of modern science, this issue has come to mean a matter of life and death for civilization as we know it; nevertheless, for all the zeal displayed, every attempt at its solution has ended in a lamentable breakdown.

To Sigmund Freud, July 30, 1932. Published, along with Freud's reply, as *Why War?* by the League of Nations. Einstein Archive 32-543; also quoted in Nathan and Norden, *Einstein on Peace*, 188

Anyone who really wants to abolish war must resolutely declare himself in favor of his own country's resigning a portion of its sovereignty in place of international institutions.

From "America and the Disarmament Conference of 1932." Published in *Mein Weltbild*, 63; reprinted in *Ideas and Opinions*, 101

As long as armies exist, any serious conflict will lead to war. A pacifism that does not actively fight against the armament of nations is and must remain impotent.

From "Active Pacifism." Published in *Mein Weltbild* (1934), 55; reprinted in *Ideas and Opinions*, 111

[The likelihood of transforming matter into energy] is something akin to shooting birds in the dark in a country where there are only a few birds.

> Remark at a 1935 press conference, three years before the atom was successfully split to cause fission. Quoted in Nathan and Norden, *Einstein on Peace*, 290

It is unworthy of a great nation to stand idly by while small countries of great culture are being destroyed with a cynical contempt for justice.

> From a message to a peace meeting, April 5, 1938. Ibid., 279

Organized power can be opposed only by organized power. Much as I regret this, there is no other way.

> To a pacifist student, July 14, 1941. Quoted in Nathan and Norden, *Einstein on Peace*, 319

I have done no work on [the atomic bomb], no work at all. I am interested in the bomb the same as any other person, perhaps a little bit more interested.

> From an interview with Richard J. Lewis, *New York Times*, August 12, 1945

As long as nations demand unrestricted sovereignty we shall undoubtedly be faced with still bigger wars, fought with bigger and technologically more advanced weapons.

To Robert Hutchins, September 10, 1945. Quoted in Nathan and Norden, *Einstein on Peace*, 337

The release of atomic energy has not created a new problem. It has merely made more urgent the necessary solving of an existing one. One could say it has affected us quantitatively, not qualitatively.

From "Atomic War or Peace," *Atlantic Monthly*, November 1945

I do not believe that civilization will be wiped out in a war fought with the atomic bomb. Perhaps two-thirds of the people on earth would be killed, but enough men capable of thinking, and enough books, would be left to start out again, and civilization would be restored.

Ibid.

The secret of the bomb should be committed to a world government. . . . Do I fear the tyranny of a world government? Of course I do. But I fear still more the coming of another war or wars. Any government is certain to be evil to some extent. But a world government is preferable to the far greater evil of wars.

Ibid.

I do not consider myself the father of the release of atomic energy. My part in it was quite indirect. I did not, in fact, foresee that it would be released in my lifetime. I believed only that it was theoretically possible. It became practical only through the accidental discovery of a chain reaction, and this was not something I could have predicted.

Ibid.

It should not be forgotten that the atomic bomb was made in this country as a preventive measure; it was to head off its use by the Germans if they discovered it.

Ibid.

I am not saying the U.S. should manufacture and stockpile the bomb, for I believe that it must do so; it must be able to deter another nation from making an atomic attack.

Ibid.

Since I do not foresee that atomic energy is to be a great boon for a long time, I have to say that for the present it is a menace. Perhaps it is good that it is so. It may intimidate the human race into bringing order into its international affairs, which, without the pressure of fear, it would not do.

Ibid.

The war is won, but the peace is not.

Statement at the fifth Nobel Anniversary Dinner in
New York. Published in the *New York Times*,
December 11, 1945; reprinted in *Ideas and Opinions*,
115–117

* Past thinking and methods did not prevent world
wars. Future thinking must.

Ibid.

* Bullets kill men, but atomic bombs kill cities. A tank
is a defense against a bullet, but there is no defense
against a weapon that can destroy civilization. . . .
Our defense is in law and order.

Ibid.

* Science has brought forth this danger, but the real
problem is in the minds and hearts of men. We will
change the hearts of other men [only] by changing
our own hearts and speaking bravely.

Ibid.

Noncooperation in military matters should be an
essential moral principle for all true scientists . . .
who are engaged in basic research.

In answer to a question posed by the Overseas
News Agency, January 20, 1947. Quoted in Nathan
and Norden, *Einstein on Peace*, 401

Had I known that the Germans would not succeed in producing an atomic bomb, I never would have lifted a finger.

To *Newsweek* magazine, March 10, 1947, in regard to sending the famous letter to President Roosevelt about the new possibility of constructing atom bombs. See Appendix for complete text of the letter. Einstein also maintained that the development of nuclear energy would have proceeded much the same even without his intervention.

* Through the release of atomic energy, our generation has brought into the world the most revolutionary force since prehistoric man's discovery of fire.

Letter written on behalf of the Emergency Committee of Atomic Scientists, March 22, 1947. Einstein Archive 70-918

It is characteristic of the military mentality that nonhuman factors (atom bombs, strategic bases, weapons of all sorts, the possession of raw materials, etc.) are held essential, while the human being, his desires and thoughts—in short, the psychological factors—are considered unimportant and secondary. . . . The individual is degraded . . . to "human materiel."

From "The Military Mentality," *American Scholar*, Summer 1947

* The bombing of civilian centers was initiated by the Germans and adopted by the Japanese. To it, the Allies responded in kind—as it turned out, with greater effectiveness—and they were morally justified in doing so.

> From an interview with Raymond Swing, *Atlantic Monthly*, November 1947

* Deterrence should be the only purpose of the stockpile of bombs.... To keep a stockpile of atomic bombs without promising not to initiate their use is to exploit possession of the bombs for political ends.... [Otherwise] atomic warfare will be hard to avoid.

> Ibid.

* A strength of the communist system of the East is that it has some of the character of a religion and inspires the emotions of a religion. Unless the force of peace, based on law, gathers behind it the force and zeal of religion, it can hardly hope to succeed.... There must be added that deep power of emotion that is a basic ingredient of religion.

> Ibid.

As long as there is man, there will be war.

> To Philippe Halsman, 1947. Einstein Archive
> 58-260; also quoted on p. 35 of *Time*, December 31,
> 1999, in its "Man of the Century" coverage of
> Einstein

Where belief in the omnipotence of physical force
gets the upper hand in political life, this force takes
on a life of its own and proves stronger than the
men who think to use force as a tool.

> From an address at Carnegie Hall in New York on
> receiving the One World Award, April 27, 1948.
> Published in *Out of My Later Years*; reprinted in
> *Ideas and Opinions*, 147

We scientists, whose tragic destiny it has been to
help make the methods of annihilation ever more
gruesome and more effective, must consider it
our solemn and transcendent duty to do all in
our power to prevent these weapons from being
used for the brutal purpose for which they were
invented.

> Quoted in the *New York Times*, August 29, 1948

* Although the progress achieved by physicists may
lead to ways of applying it in a technical and military way that might involve extreme dangers, the
responsibility lies with those who are *using* the
means and not with those who lead in the progress

toward knowledge—that is, with the politicians, not with the scientists.

> Answer to a questionnaire asking if the scientists
> who developed the bomb should be held
> responsible for its destructive outcome, October
> 1948. See Kaller's Autographs catalog, "Jewish
> Visionaries," 37. Einstein Archive 58-015

* The creation of a United States of Europe is a necessity if one considers the economic and technical situation. Whether this union will mean a secure peace cannot be predicted by anyone with certainty, but I think a "yes" is more likely than a "no."

> In answer to the question of whether such an
> alliance would solve the problem of war. Ibid.

Responsibility lies with those who make use of these new tools and not with those who contribute to the progress of knowledge: therefore, with the politicians, not with the scientists.

> From an interview by student Milton James,
> February 1949. Einstein Archive 58-014

* I do not know how the Third World War will be fought, but I can tell you what they will use in the Fourth—rocks!

> From an interview with Alfred Werner, *Liberal
> Judaism* 16 (April–May 1949), 12

So long as security is sought through national armament, no country is likely to renounce any weapon that seems to promise it victory in the event of war. In my opinion, security can be attained only by renouncing all national military defense.

To Jacques Hadamard, December 29, 1949.
Einstein Archive 12-064

If it [the effort to produce a hydrogen bomb] is successful, radioactive poisoning of the atmosphere and hence annihilation of any life on earth will have been brought within the range of what is technically possible.

From a contribution to Eleanor Roosevelt's
television program on the implications of the
H-bomb, February 13, 1950. Published in *Ideas
and Opinions*, 159–161

* Competitive armament is not a way to prevent war. Every step in this direction brings us nearer to catastrophe. . . . I repeat, armament is no protection against war, but leads inevitably *to* war.

From a United Nations Radio interview, June 16,
1950, recorded in the study of Einstein's home in
Princeton. Published in *Ideas and Opinions*, 161–163

* Striving for peace and preparing for war are incompatible with each other. . . . Arms must be entrusted only to an international authority.

> Ibid.

* One cannot abolish only a single weapon, but only war as a whole.

> From a draft manuscript, ca. 1950, on the
> formation of world government. See Kaller's
> Autographs catalog, "Jewish Visionaries," 38

To my mind, to kill in war is not a whit better than to commit ordinary murder.

> Quoted in the Japanese magazine *Kaizo*, Autumn
> 1952

The first atomic bomb destroyed more than the city of Hiroshima. It also exploded our inherited, outdated political ideas.

> A co-signed statement quoted in the *New York
> Times*, June 12, 1953

There was never even the slightest indication of any potential technological application.

> To Jules Isaac, February 28, 1955, refuting the idea
> that his special theory of relativity was responsible
> for atomic fission and the atom bomb. Atomic
> fission, accomplished in December 1938 in Berlin

by Otto Hahn and Fritz Strassmann, was made
possible by the discovery of the neutron by James
Chadwick in 1932; fission requires neutrons.
Quoted in Nathan and Norden, *Einstein on Peace*,
623

There lies before us, if we choose, continued progress in happiness, knowledge, and wisdom. Shall
we, instead, choose death, because we cannot forget our quarrels? We appeal, as human beings, to
human beings: Remember your humanity and forget the rest.

Einstein's last signed statement, issued with
Bertrand Russell, April 11, 1955, one week before
Einstein's death. Einstein Archive 33-212

The unleashing of power of the atom bomb has
changed everything except our mode of thinking,
and thus we head toward unparalleled catas-
trophes.

Quoted in the *New York Times Magazine*, August 2,
1964

I made one mistake in my life—when I signed that
letter to President Roosevelt advocating that the
bomb should be built. But perhaps I can be forgiven
for that because we all felt that there was a high
probability that the Germans were working on this

problem and they might succeed and use the atomic bomb to become the master race.

To Linus Pauling, recorded in Pauling's diary. See Clark, *Einstein*, 554, 847n; also quoted by Ted Morgan in *FDR* (New York: Simon and Schuster, 1985), and by Thomas Hager in *Force of Nature: The Life of Linus Pauling* (New York: Simon and Schuster, 1995), 451, 669n

On Politics, Patriotism, and Government

Einstein in 1933. (Courtesy of the Archives, California Institute of Technology)

* The state to which I belong as a citizen plays not the slightest role in my personal life. I regard a person's relationship with the state as a business matter, akin to one's relationship to life insurance.

A passage deleted from "My Opinion on the War," a manuscript for the Berliner Goethebund, published in 1916. *CPAE*, Vol. 6, Doc. 20

Nationalism is an infantile disease. It is the measles of mankind.

Statement to G. S. Viereck, 1921. Quoted in Dukas and Hoffmann, *Albert Einstein, the Human Side*, 38

* I wish (1) that next year will bring the broadest possible international agreements on disarmament on land and at sea; (2) that a solution will be found for the international war debts that allows the European states to pay their obligations without having to pawn their property abroad; (3) that an honest arrangement can be reached with the Soviet Union that frees this land from external pressures while allowing its internal development to proceed unhindered.

Statement for the December 31, 1928, *Chicago Daily News*, after being asked by reporter Edgar Mowrer what his wishes were for the New Year. It was scribbled in German on the back of Mowrer's letter of December 18. (Courtesy of Uriel Gorney and Mishael Zedek.)

My political ideal is that of democracy. Let every man be respected as an individual and no man be idolized.

From "What I Believe," *Forum and Century* 84 (1930),
193–194; reprinted in *Ideas and Opinions*, 8–11

The state is made for man, not man for the state. . . .
That is to say, the state should be our servant and not we its slaves.

From "The Disarmament Conference of 1932," *The
Nation* 33 (1931), 300; reprinted in *Mein Weltbild*,
57, and in *Ideas and Opinions*, 95. In this article,
Einstein conceded that "these are old sayings."

* Hitler is living on the empty stomach of Germany. . . . An empty stomach is not a good political adviser . . . [but] better political insight has a hard time winning as long as there is little prospect of filling the stomach.

In *Cosmic Religion* (1931), 107. (Thanks to young
Brian Claeys of Iowa for this source.)

As long as I have any choice in the matter, I will live only in a country where civil liberty, tolerance, and equality of all citizens before the law are the rule. . . . These conditions do not exist in Germany at the present time.

From "Manifesto," March 1933. Published in *Mein
Weltbild*, 81; reprinted in *Ideas and Opinions*, 205

Nationalism is, in my opinion, nothing more than an idealistic rationalization for militarism and aggression.

From the first draft of a speech at Royal Albert Hall, London, October 3, 1933. Quoted in Nathan and Norden, *Einstein on Peace*, 242

* National loyalty is limiting; men must be taught to think in world terms. Every country will have to surrender a portion of its sovereignty through international cooperation. To avoid destruction, aggression must be sacrificed.

From an interview, *Survey Graphic* 24 (August 1935), 384, 413

Politics is a pendulum whose swings between anarchy and tyranny are fueled by perennially rejuvenated illusions.

Aphorism, 1937. Quoted in Dukas and Hoffmann, *Albert Einstein, the Human Side*, 38

There are times when the climate of the world is good for ethical things. Sometimes men trust one another and create good. At other times, it is not so.

From a conversation recorded by Algernon Black, Fall 1940. Einstein Archive 54-834

When people live in a time of maladjustment, when there is tension and disequilibrium, they become unbalanced themselves and then may follow an unbalanced leader.

> Ibid.

The greatest weakness of the democracies is economic fear.

> Ibid.

* I am convinced that an international political organization is not only possible but is unconditionally necessary if the situation on our planet should eventually become unbearable.

> From a draft manuscript, ca. 1940. See Kaller's Autographs catalog, "Jewish Visionaries," 35

* There is no other salvation for civilization and even for the human race than in the creation of a world government, with the security of nations founded upon law. As long as there are sovereign states with their separate armaments and armament secrets, new world wars cannot be avoided.

> *New York Times*, September 15, 1945

Everything that is done in international affairs must be done from the following viewpoint: Will it help or hinder the establishment of world government?

From the text of a broadcast interview with P. A. Schilpp and F. Parmelee, May 29, 1946. Einstein Archive 29-105; see also Nathan and Norden, *Einstein on Peace*, 382

A world government must be created which is able to solve conflicts between nations by judicial decision. . . . This government must be based on a clear-cut constitution that is approved by the governments and the nations, and which has the sole disposition of offensive weapons.

New York Times, May 30, 1946. Quoted in Pais, *Einstein Lived Here*, 232

We must learn the difficult lesson that the future of mankind will only be tolerable when our course, in world affairs as in all other matters, is based upon justice and law rather than the threat of naked power.

From a message for the Gandhi memorial service, February 11, 1948. Quoted in Nathan and Norden, *Einstein on Peace*, 468

There is only *one* path to peace and security: the path of a supranational organization. One-sided armament on a national basis only heightens the

general uncertainty and confusion without being
an effective protection.

From an address at Carnegie Hall in New York on
receiving the One World Award, April 27, 1948.
Published in *Out of My Later Years*; reprinted in
Ideas and Opinions, 147

I advocate world government because I am convinced that there is no other possible way of elimi-
nating the most terrible danger in which man has
ever found himself. The objective of avoiding total
destruction must have priority over any other
objective.

Reply to a Soviet scientist's open letter in the *New
York Times*, 1948. Quoted in *Einstein on Humanism*,
45

To act intelligently in human affairs is only possible
if an attempt is made to understand the thoughts,
motives, and apprehensions of one's opponent so
fully that one can see the world through his eyes.

Ibid., 39

Considerable economic security at the expense of
liberty and political rights.

Describing communism, October 7, 1948, in
answer to a question posed by Milton James.
Einstein Archive 58-015

If the idea of world government is not realistic, then there is only *one* realistic view of our future: wholesale destruction of man by man.

Comment about the film *Where Will You Hide?* 1948. Einstein Archive 28-817

I have never been a Communist. But if I were, I would not be ashamed of it.

To Lydia B. Hewes, July 10, 1950. Einstein Archive 59-984

Mankind can be saved only if a supranational system, based on law, is created to eliminate the methods of brute force.

Statement in *Impact* 1 (1950), 104

I can see only the revolutionary way of non-cooperation in the sense of Gandhi's. Every intellectual who is called before one of the committees ought to refuse to testify; i.e., he must be prepared for jail and economic ruin . . . in the interest of the cultural welfare of his country.

To Brooklyn teacher William Frauenglass, May 16, 1953, who was called before Senator Joseph McCarthy's House Un-American Activities Committee hearings. Einstein Archive 41-112

* Refusal to testify must be based on the assertion that it is shameful for a blameless citizen to submit to such an inquisition and that this kind of inquisition violates the spirit of the Constitution.

> Ibid.

There is no such [anti-Communist] hysteria in the West European countries and there is no danger of their governments being overthrown by force or subversion, despite the fact that Communist parties are not persecuted or even ostracized.

> To E. Lindsay, July 18, 1953. Einstein Archive 60-326

Eastern Europe would never have become prey to Russia if the Western powers had prevented German aggressive fascism under Hitler, which grave mistake made it necessary afterwards to beg Russia for help.

> Ibid.

Party membership is a thing for which no citizen is obligated to give an accounting.

> To C. Lamont, January 2, 1954. Einstein Archive 60-178

The fear of communism has led to practices that have become incomprehensible to the rest of civilized mankind and expose our country to ridicule.

Message to the Decalogue Society of Lawyers
on receiving its merit award. *New York Times*,
February 21, 1954

The current [House Un-American Activities Committee] investigations are an incomparably greater danger to our society than those few Communists in the country ever could be. These investigations have already undermined to a considerable extent the democratic character of our society.

To Felix Arnold, March 19, 1954. Einstein Archive
59-118

In Plato's time, and even later, in Jefferson's time, it was still possible to reconcile democracy with a moral and intellectual aristocracy, while today democracy is based on a different principle—namely, that the other fellow is not better than I am.

On democracy and anti-intellectualism, in
Niccolo Tucci's *New Yorker* profile of Einstein,
November 22, 1954

Political passions, aroused everywhere, demand their victims.

Final written words, in an unpublished
manuscript, probably April 12–14, 1955. Quoted in
Pais, *Subtle Is the Lord*, 530

That is simple, my friend: because politics is more difficult than physics.

> When asked why people could discover atoms but not the means to control them. Recalled in the *New York Times*, April 22, 1955, after his death

In my opinion it is not right to bring politics into scientific matters, nor should individuals be held responsible for the government of the country to which they happen to belong.

> To H. A. Lorentz. Quoted in French, *Einstein: A Centenary Volume*, 187

One must divide one's time between politics and equations. But our equations are much more important to me, because politics is for the present, while our equations are for eternity.

> Said to Ernst Straus. Quoted in Seelig, *Helle Zeit, dunkle Zeit*, 71

Einstein, the "Jewish saint," embedded among saints, angels, and other worthy souls in the tympanum of the entrance to New York City's Riverside Church, built 1927–1930. The Reverend Harry E. Fosdick, the church's first senior minister, explained that Einstein belonged there because "he that doeth good is of God," a New Testament description of the saintly. Einstein is said to have been delighted by the image. (Photo by Alice Calaprice)

Einstein's "religion," as he often explained it, was an attitude of cosmic awe and wonder and a devout humility before the harmony of nature, rather than a belief in a personal God who is able to control the lives of individuals. He referred to this "belief" as "cosmic religion." It is incompatible with the doctrines of all theistic religions in its denial of a personal God who punishes the wicked and rewards the righteous. See Jammer, *Einstein and Religion*, 149

Why do you write to me, "God should punish the English"? I have no close connection to either one or the other. I see only with deep regret that God punishes so many of his children for their numerous stupidities, for which he himself can be held responsible; in my opinion, only his nonexistence could excuse him.

To Edgar Meyer, a Swiss colleague, January 2, 1915. *CPAE*, Vol. 8, Doc. 44

* Upon reading books on philosophy, I learned that I stood there like a blind man in front of a painting. I can grasp only the inductive method . . . the works of speculative philosophy are beyond my reach.

To Eduard Hartmann, April 27, 1917. *CPAE*, Vol. 8, Doc. 330

* I want to know how God created this world. I am not interested in this or that phenomenon, in the spectrum of this or that element. I want to know his thoughts. The rest are details.

> Said to his Berlin student Esther Salaman,
> probably around 1920. See Salaman, "A Talk with
> Einstein," *Listener* 54 (1955), 370–371

In every true searcher of Nature there is a kind of religious reverence, for he finds it impossible to imagine that he is the first to have thought out the exceedingly delicate threads that connect his perceptions.

> 1920. In Moszkowski, *Conversations with Einstein*,
> 46

Since our inner experiences consist of reproductions and combinations of sensory impressions, the concept of a soul without a body seems to me to be empty and devoid of meaning.

> To a Viennese woman, February 5, 1921. Einstein
> Archive 43-847; also quoted in Dukas and
> Hoffmann, *Albert Einstein, the Human Side*, 40

* The sense of the word "truth" can vary according to its uses as a life experience, a mathematical theorem, or a theory in natural science. I do not think there is anything clear-cut about a "religious truth."

In answer to the question, Do scientific and religious truths come from different points of view?
posed by interviewers for *Kaizo* 5, no. 2 (1923), 197

* Scientific research, through the fostering of causa-
tive thinking and evaluation, can repudiate superstition. For sure, a conviction in the reasonableness
and comprehensibility of the world, in kinship with
religious feelings, is at the basis of all the most ele-
gant scientific work.

In answer to the question, Can scientific discovery
enhance religious belief and repudiate
superstition, since religious feelings can give
impetus to scientific discovery? Ibid.

* My comprehension of God comes from the deeply
felt conviction of a superior intelligence that reveals
itself in the knowable world. In common terms, one
can describe it as "pantheistic" (Spinoza).

In answer to the question, What is your
understanding of God? Ibid.

* I can look at doctrinaire traditions only with a historical and psychological perspective; I have no
other connection with them.

In answer to the question, What is your opinion
regarding a "savior"? Ibid.

I cannot conceive of a personal God who would directly influence the actions of individuals. . . . My religiosity consists of a humble admiration of the infinitely superior spirit that reveals itself in the little that we can comprehend of the knowable world. That deeply emotional conviction of the presence of a superior reasoning power, which is revealed in the incomprehensible universe, forms my idea of God.

> To a banker in Colorado, August 1927. Einstein
> Archive 48-380; also quoted in Dukas and
> Hoffmann, *Albert Einstein, the Human Side*, 66, and
> in the *New York Times* obituary, April 19, 1955

Everything is determined . . . by forces over which we have no control. It is determined for the insect as well as for the star. Human beings, vegetables, or cosmic dust—we all dance to a mysterious tune, intoned in the distance by an invisible piper.

> In the *Saturday Evening Post*, October 26, 1929.
> Quoted in Clark, *Einstein*, 422

I believe in Spinoza's God who reveals himself in the harmony of all that exists, but not in a God who concerns himself with the fate and actions of human beings.

> Telegram to a Jewish newspaper, 1929. Einstein
> Archive 33-272. (Spinoza reasoned that God and
> the material world are indistinguishable; the better

one understands how the universe works, the
closer one comes to God.)

* There are two different conceptions about the nature of the universe: (1) the world as a unity dependent on humanity; (2) the world as a reality independent of the human factor.

  > From a conversation with Indian mystic, poet, and
  > musician Rabindranath Tagore, July 14, 1930.
  > Published in Tagore, *The Religion of Man* (New
  > York: Macmillan, 1931), Appendix 2, 221–225

* I cannot *prove* scientifically that Truth must be conceived as a truth that is valid independent of humanity, but I firmly believe it. . . . If there is a reality
  independent of man, there is also a truth relative to
  this reality. . . . The problem begins with whether
  Truth is independent of our consciousness. . . . For
  instance, if nobody is in this house, that table remains where it is.

  > Ibid.

The man who is thoroughly convinced of the universal operation of the law of causation cannot for
a moment entertain the idea of a being who inter-
feres in the course of events. . . . He has no use for
the religion of fear and equally little for social or
moral religion. A God who rewards and punishes
is inconceivable to him for the simple reason that a

man's actions are determined by necessity, external and internal, so that in God's eyes he cannot be responsible, any more than an inanimate object is responsible for the motions it undergoes.... A man's ethical behavior should be based effectively on sympathy, education, and social relationships; no religious basis is necessary. Man would indeed be in a poor way if he had to be restrained by fear of punishment and hope of reward after death.

> From "Religion and Science," in the *New York Times Magazine*, November 9, 1930, 1–4. Also in *Cosmic Religion*, 51–52. In German in *Berliner Tageblatt*, November 11, 1930

Everything that the human race has done and thought is concerned with the satisfaction of deeply felt needs and the assuagement of pain. One has to keep this constantly in mind if one wishes to understand spiritual movements and their development. Feeling and longing are the motive forces behind all human endeavors and human creations.

> Ibid.

I am of the opinion that all the finer speculations in the realm of science spring from a deep religious feeling.... I also believe that this kind of religiousness ... is the only creative religious activity of our time.

> In *Forum and Century* 83 (1930), 373

I cannot conceive of a God who rewards and punishes his creatures, or has a will of the kind that we experience in ourselves. Neither can I nor would I want to conceive of an individual who survives his physical death; let feeble souls, from fear or absurd egoism, cherish such thoughts.

From "What I Believe," *Forum and Century* 84
(1930), 193–194; reprinted in *Ideas and Opinions*, 11

* We know nothing about it [God, the world] at all. All our knowledge is but the knowledge of schoolchildren. Possibly we shall know a little more than we do now. But the real nature of things, that we shall never know, never.

From an interview with Chaim Tschernowitz, *The Jewish Sentinel*, September 1931

It is very difficult to elucidate this [cosmic religious] feeling to anyone who is entirely without it. . . . The religious geniuses of all ages have been distinguished by this kind of religious feeling, which knows no dogma and no God conceived in man's image; so that there can be no church whose central teachings are based on it. . . . In my view, it is the most important function of art and science to awaken this feeling and keep it alive in those who are receptive to it.

On "cosmic religion," a worship of the harmony and beauties of nature that became the common faith of physicists. In *Cosmic Religion* (1931), 48–49

* I will call it the cosmic religious sense. This is hard to make clear to those who do not experience it, since it does not involve an anthropomorphic idea of God; the individual feels the vanity of human desires and aims, and the nobility and marvelous order which are revealed in nature and in the world of thought.

> Ibid., 48

I assert that the cosmic religious experience is the strongest and the noblest driving force behind scientific research.

> Ibid., 52

* I see a pattern, but my imagination cannot picture the maker of that pattern. I see a clock, but I cannot envision the clockmaker. The human mind is unable to conceive of the four dimensions, so how can it conceive of a God, before whom a thousand years and a thousand dimensions are as one?

> Ibid., 102

Philosophy is like a mother who gave birth to and endowed all the other sciences. Therefore one should not see her in her nakedness and poverty, but should hope, rather, that part of her Don Quixote ideal will live on in her children so that they do not sink into philistinism.

To Bruno Winawer, September 8, 1932. Einstein
Archive 36-532; also quoted in Dukas and
Hoffmann, *Albert Einstein, the Human Side*,
106

Our actions should be based on the ever-present
awareness that human beings in their thinking,
feeling, and acting are not free but are just as caus-
ally bound as the stars in their motion.

Statement to the Spinoza Society of America,
September 22, 1932. Einstein Archive 33-291

I cannot imagine a God who rewards and punishes
the objects of his creation, whose purposes are
modeled after our own—a God, in short, who is but
a reflection of human frailty. . . . It is enough for me
to contemplate the mystery of conscious life perpet-
uating itself through all eternity, to reflect upon the
marvelous structure of the universe which we can
dimly perceive and to try humbly to comprehend
even an infinitesimal part of the intelligence mani-
fested in Nature.

From "My Credo," for the German League of
Human Rights, 1932. Quoted in Leach, *Living
Philosophies*, 3

* If one purges all subsequent additions from the
original teachings of the Prophets and Christianity,
especially those of the priests, one is left with a

pedagogy that is capable of curing all the social ills of humankind.

Statement for the Romanian Jewish journal *Renasterea Noastra*, January 1933. Published in *Mein Weltbild*; reprinted in *Ideas and Opinions*, 184–185

Organized religion may regain some of the respect it lost in the last war if it dedicates itself to mobilizing the goodwill and energy of its followers against the rising tide of illiberalism.

New York Times, April 30, 1934. Also quoted in Pais, *Einstein Lived Here*, 205

You will hardly find one among the profounder sort of scientific minds without a religious feeling of his own. But it is different from the religiosity of the naïve man. For the latter, God is a being from whose care one hopes to benefit and whose punishment one fears; a sublimation of a feeling similar to that of a child for its father.

From "The Religious Spirit of Science." Published in *Mein Weltbild* (1934), 18; reprinted in *Ideas and Opinions*, 40

The scientist is possessed by a sense of universal causation. . . . His religious feeling takes the form of a rapturous amazement at the harmony of natural law, which reveals an intelligence of such superiority that, compared with it, all the systematic thinking and acting of human beings is an utterly insignificant reflection. . . . It is beyond question closely akin to that which has possessed the religious geniuses of all ages.

Ibid.

What is the meaning of human life, or for that matter, of the life of any creature? To know an answer to this question means to be religious. You ask: Does it make any sense, then, to pose this question? I answer: The man who regards his own life and that of his fellow creatures as meaningless is not merely unhappy but hardly fit for life.

Published in *Mein Weltbild* (1934), 10; reprinted in *Ideas and Opinions*, 11

* Everyone has been given an endowment that he must strive to develop in the service of mankind. This cannot be brought to completion through the threat of a God who will punish man for sin, but only by challenging the best in human nature.

From an interview, *Survey Graphic* 24 (August 1935), 384, 413

Everyone who is seriously involved in the pursuit of science becomes convinced that a spirit is manifest in the laws of the universe—a spirit vastly superior to that of man. . . . In this way the pursuit of science leads to a religious feeling of a special sort, which is indeed quite different from the religiosity of someone more naïve.

To student Phyllis Wright, who asked if scientists pray, January 24, 1936. Einstein Archive 42-601, 52-337

Whatever there is of God and goodness in the universe, it must work itself out and express itself through us. We cannot stand aside and let God do it.

From a conversation recorded by Algernon Black, Fall 1940. Einstein Archive 54-834

A religious person is devout in the sense that he has no doubt about the significance of those super-personal objects and goals that neither require nor are capable of rational foundation.

Nature 146 (1940), 605

To [the sphere of religion] belongs the faith in the possibility that the regulations valid for the world of existence are rational, that is, comprehensible to reason. I cannot conceive of a genuine scientist without that profound faith.

From "Science, Philosophy, and Religion," a written contribution to a symposium held in New York in 1940 on how science, philosophy, and religion advance the cause of American

democracy; published in 1941. Einstein Archive
28-523; reprinted in *Ideas and Opinions* as "Science
and Religion," 46

Science without religion is lame, religion without
science is blind.

Ibid. See *Ideas and Opinions*, 46. This may be a play
on Kant's "Notion without intuition is empty,
intuition without notion is blind"; Einstein was not
always totally original. Some scientists, perhaps
many, disagree with Einstein's sentiment. (See, for
example, Dyson, "Writing a Foreword for Alice
Calaprice's New Einstein Book," 491–502.)

The main source of the present-day conflicts between the spheres of religion and science lies in the
concept of a personal God.

Ibid. See *Ideas and Opinions*, 47

In their struggle for the ethical good, teachers of religion must have the stature to give up the doctrine
of a personal God, that is, give up that source of
fear and hope which in the past placed such vast
power in the hands of priests.

Ibid., 48

The further the spiritual evolution of mankind advances, the more certain it seems to me that the
path to genuine religiosity does not lie through the

fear of life, and the fear of death, and blind faith, but through striving after rational knowledge.

Ibid., 49

* In view of such harmony in the cosmos which I, with my limited human mind, am able to recognize, there are yet people who say there is no God. But what makes me really angry is that they quote me for support of such views.

> Said to German anti-Nazi diplomat and author Hubertus zu Löwenstein around 1941. Quoted in his book, *Towards the Further Shore* (London, 1968), 156. With this remark, Einstein dissociates himself from atheism; see Jammer, *Einstein and Religion*, 97

* Then there are the fanatical atheists whose intolerance is the same as that of the religious fanatics, and it springs from the same source. . . . They are creatures who can't hear the music of the spheres.

> To an unidentified person, August 7, 1941, on the reaction to his symposium contribution, "Science, Philosophy, and Religion" (1940). Einstein Archive 54-927. To many readers, Einstein's denial of a "personal" God meant a total denial of God, because "there is no other God but a personal God." See discussion in Jammer, *Einstein and Religion*, 92–108

It is quite possible that we can do greater things than Jesus, for what is written in the Bible about him is poetically embellished.

Quoted in W. Hermanns, "A Talk with Einstein,"
October 1943. Einstein Archive 55-285

No idea is conceived in our mind independent of our five senses [i.e., no idea is divinely inspired].

Ibid.

I would not think that philosophy and reason themselves will be man's guide in the foreseeable future; however, they will remain the most beautiful sanctuary they have always been for the select few.

To Benedetto Croce, June 7, 1944. Einstein Archive
34-075; also quoted in Pais, *Einstein Lived Here*, 122

I often read the Bible, but its original text has remained beyond my reach.

To H. Friedmann, September 2, 1945, regarding his
lack of knowledge of the Hebrew language.
Quoted in Pais, *Subtle Is the Lord*, 38

It is this . . . symbolic content of the religious traditions which is likely to come into conflict with science. . . . Thus it is of vital importance for the preservation of true religion that such conflicts be avoided when they arise from subjects which, in fact, are not really essential for the pursuit of religious aims.

Statement to the Liberal Ministers Club, New York
City. Published in the *Christian Register*, June 1948

My position concerning God is that of an agnostic. I am convinced that a vivid consciousness of the primary importance of moral principles for the betterment and ennoblement of life does not need the idea of a law-giver, especially a law-giver who works on the basis of reward and punishment.

To M. Berkowitz, October 25, 1950. Einstein
Archive 59-215

I have found no better expression than "religious" for confidence in the rational nature of reality, insofar as it is accessible to human reason. Whenever this feeling is absent, science degenerates into uninspired empiricism.

To Maurice Solovine, January 1, 1951. Einstein
Archive 21-474; published in *Letters to Solovine*, 119

Mere unbelief in a personal God is no philosophy at all.

To V. T. Aaltonen, May 7, 1952, on his opinion that
belief in a personal God is better than atheism.
Einstein Archive 59-059

My feeling is religious insofar as I am imbued with the consciousness of the insufficiency of the human mind to understand more deeply the harmony of the universe which we try to formulate as "laws of nature."

> To Beatrice Frohlich, December 17, 1952. Einstein Archive 59-797

The idea of a personal God is quite alien to me and seems even naïve.

> Ibid.

To assume the existence of an imperceivable being . . . does not facilitate understanding the orderliness we find in the perceivable world.

> To an Iowa student who asked, What is God? July 1953. Einstein Archive 59-085

I do not believe in the immortality of the individual, and I consider ethics to be an exclusively human concern with no superhuman authority behind it.

> July 1953. Einstein Archive 36-553; also quoted in Dukas and Hoffmann, *Albert Einstein, the Human Side*, 39

If God created the world, his primary concern was certainly not to make its understanding easy for us.

> To David Bohm, February 10, 1954. Einstein
> Archive 8-041

I consider the Society of Friends the religious community that has the highest moral standards. As far as I know, they have never made evil compromises and are always guided by their conscience. In international life, especially, their influence seems to me very beneficial and effective.

> To A. Chapple, Australia, February 23, 1954.
> Einstein Archive 59-405; also quoted in Nathan
> and Norden, *Einstein on Peace*, 511

I do not believe in a personal God and I have never denied this but have expressed it clearly. If something is in me that can be called religious, then it is the unbounded admiration for the structure of the world so far as science can reveal it.

> To an admirer, March 22, 1954. Einstein Archive
> 39-525; also quoted in Dukas and Hoffmann, *Albert
> Einstein, the Human Side*, 43

I am a deeply religious nonbeliever. . . . This is a somewhat new kind of religion.

> To Hans Muehsam, March 30, 1954. Einstein
> Archive 38-434

I don't try to imagine a God; it suffices to stand in awe of the structure of the world, insofar as it allows our inadequate senses to appreciate it.

To S. Flesch, April 16, 1954. Einstein Archive
30-1154

I have never imputed to Nature a purpose or a goal, or anything that could be understood as anthropomorphic. What I see in Nature is a magnificent structure that we can comprehend only very imperfectly, and that must fill a thinking person with a feeling of humility. This is a genuinely religious feeling that has nothing to do with mysticism.

1954 or 1955. Quoted in Dukas and Hoffmann,
Albert Einstein, the Human Side, 39

A man's moral worth is not measured by what his religious beliefs are, but rather by what emotional impulses he has received from Nature during his lifetime.

To Sister Margrit Goehner, February 1955. Einstein
Archive 59-831

Thus I came . . . to a deep religiosity, which, however, reached an abrupt end at the age of twelve. Through the reading of popular scientific books I soon reached a conviction that much in the stories of the Bible could not be true. . . . Suspicion against

every kind of authority grew out of this experience
. . . an attitude that has never left me.

"Autobiographical Notes," in Schilpp, *Albert
Einstein: Philosopher-Scientist*, 9

Out yonder there was this huge world, which exists
independently of us human beings and which
stands before us like a great, eternal riddle, at least
partially accessible to our inspection and thinking.
The contemplation of this world beckoned like a
liberation, and I soon noticed that many a man I
had learned to esteem and admire had found inner
freedom and security in devoted occupation with it.

Ibid., 95

Isn't all of philosophy as if written in honey? Something may appear clear at first, but when one looks
again it has disappeared. Only the pap remains.

Quoted in Rosenthal-Schneider, *Reality and
Scientific Truth*, 90

My views are near those of Spinoza: admiration for
the beauty and belief in the logical simplicity of the
order and harmony that we can grasp humbly
and only imperfectly. I believe that we have to content ourselves with our imperfect knowledge and

understanding and treat values and moral obligations as purely human problems.

Quoted in Hoffmann, *Albert Einstein: Creator and Rebel*, 95

* What really interests me is whether God could have created the world any differently; in other words, whether the demand for logical simplicity leaves any freedom at all.

Said to Ernst Straus, on the question of whether God had any choice in the design of the world. Quoted in Seelig, *Helle Zeit, dunkle Zeit*, 72

On Science and Scientists, Mathematics, and Technology

Unendliche, wenn es sich y dem Werte c ~
~endlichen Energie-Aufwandes, um de
~keit c zu erteilen. Um zu sehen, dass dieser
~undvielfeit in den von Newtons Mechan
war der Nenner (nach Potenzen von $\frac{q^2}{c^2}$) und erhalten

$$E\frac{1}{} = mc^2 + \frac{m}{2}q^2 + \cdots \quad (28')$$

lied — das rechten Seite ist der geläufige
aussehen Mechanik. Was bedeutet aber das
hat zwar streng genommen hier keine Leg
eine additive Konstante willkürlich weg
Blick auf 28), lass dass Glied mc^2 m
$\frac{m}{2}q^2$ untrennbar verbunden ist. Man m
le dagegen, diesem Term mc^2 eine reale Bed
des ruhenden Punktes anzusehen. N
~~Ene totale Energie~~ hätten wir
m als einen Energievorrat von der Grösse
Körpers). ~~Ist diese Auffassung zutreffend,~~
können wir aber anders, z. B. indem
mc^2 stets der Ruhe-Energie des Körpers
die träge Masse des Körpers
ser Erwärmung auch ~~~~ ändern, und zu

$E = mc^2$ written in Einstein's own hand in his manuscript of 1912 on the special theory of relativity. It shows that the E was originally an L.

Einstein is most famous for his theory of relativity. In his first papers he referred to it as the "principle of relativity." The term "theory of relativity" was first used by Max Planck in 1906 to describe the Lorentz-Einstein equations of motion for the electron, and it was finally adopted by Einstein in 1907 when he replied to an article by Paul Ehrenfest, who had also used Planck's term. But Einstein continued to use "relativity principle" in the titles of articles for several years, since a "principle" is not a "theory" but something that is borne in mind when formulating a theory. In 1910 the mathematician Felix Klein suggested "theory of invariants" as a better alternative, but "theory of relativity" stuck. In 1915 Einstein began to refer to the 1905 theory, having to do with space and time, as the "special theory," to distinguish it from his new theory of gravitation, the "general theory." See Stachel et al., *Einstein's Miraculous Year*, 101–102, and Fölsing, *Albert Einstein*, 208–210

$E = mc^2$.

Statement of the equivalence of mass and energy—energy equals mass times the speed of light squared—which opened up the atomic age. The original statement was: "If a body emits the energy L in the form of radiation, its mass decreases by L/V^2." (Originally in "Ist die Trägheit eines Körpers von seinem Energieinhalt abhängig?" *Annalen der Physik* 18 [1905], 639–641. See Stachel et al., *Einstein's Miraculous Year*, 161, for a translation of this paper.) Note that Einstein used L to represent energy until 1912, when, in his "Manuscript on the Special Theory of Relativity" (see *CPAE*, Vol. 4, Doc. 1), he crossed out the *L* and substituted *E* in equations 28 and 28' of the handwritten manuscript. (See the facsimile edition of the manuscript published by George Braziller with the Safra Foundation and the Israel Museum, Jerusalem [1996], 119 and 121; and *CPAE*, Vol. 4, Doc. 1, 58–59.) The equation derives from the special theory of relativity, which played a decisive role in the investigation and development of nuclear energy. A mass can be converted into a vast amount of energy (i.e., when a particle is released from an atom it is converted to energy), demonstrating a fundamental relationship in nature. The theory also introduced a new definition of space and time. For direct experimental evidence in its favor, however, it had to wait twenty-five years, when the conversion of mass into energy was confirmed in the study of nuclear reactions; time dilation was not directly proved until 1938. (*Note*: The following four quotations are out of chronological order, but I placed them here because they reflect Einstein's thoughts leading to the 1905 special theory.)

* What if one were to run after a ray of light? . . . What if one were riding on the beam? . . . If one were to run fast enough, would it no longer move at all? . . . What is the "velocity of light"? If it is in relation to something, this value does not hold in relation to something else which is itself in motion.

Based on a conversation with psychologist Max Wertheimer in 1916, in which Einstein attempted

to explain his thought process when he formulated
the special theory of relativity. See Wertheimer,
Productive Thinking (1945; reprinted by Harper,
1959), 218.

* After ten years of reflection, such a principle resulted from a paradox upon which I had already hit
at the age of sixteen: if I pursue a beam of light with
velocity *c* (velocity of light in a vacuum), I should
observe such a beam of light as an electromagnetic
field at rest, though spatially oscillating. . . . From
the very beginning it appeared to me intuitively
clear that, judged from the standpoint of such an
observer, everything would have to happen according to the same laws as for an observer who,
relative to earth, was at rest.

From "Autobiographical Notes," in Schilpp, *Albert
Einstein: Philosopher-Scientist*, 53

* Five or six weeks elapsed between the conception
of the idea for the special theory of relativity and
the completion of the relevant publication.

To his biographer, Carl Seelig, March 11, 1952.
Einstein Archive 39-013

* My direct path to the special theory of relativity
was mainly determined by the conviction that the
electromotive force induced in a conductor moving

in a magnetic field is nothing other than an electric field.

> For a message read at a celebration of the centennial of Albert Michelson's birth, December 19, 1952, at Case Institute. See Stachel et al., *Einstein's Miraculous Year*, 111

* According to the assumption considered here, in the propagation of a light ray emitted from a point source, the energy is not distributed continuously over ever-increasing volumes of space, but consists of a finite number of energy quanta localized at points of space that move without dividing and can be absorbed or generated only as complete units.

> From "On a Heuristic Point of View Concerning the Production and Transformation of Light," March 1905. See Stachel et al., *Einstein's Miraculous Year*, 178. Considered by some to be the most revolutionary sentence written by a twentieth-century physicist; see Fölsing, *Albert Einstein*, 143

* I've completely solved the problem. My solution was to analyze the concept of time. Time cannot be absolutely defined, and there is an inseparable relation between time and signal velocity.

> Said to Michele Besso, May 1905, in reference to his forthcoming publication, "On the Electrodynamics of Moving Bodies," on the relativity principle in electrodynamics, later to be called the special theory of relativity. Recalled during Einstein's lecture in Kyoto, December 14, 1922. See *Physics Today* (August 1982), 46

* [I will send you] four papers. [The first] deals with
radiation and the energetic properties of light and
is very revolutionary. . . . The second paper deter-
mines the true size of atoms by way of diffusion
and the viscosity of diluted solutions of neutral
substances. The third proves that, assuming the
molecular theory of heat, bodies on the order of
magnitude of 1/1000 mm, when suspended in liquids, must already have an observable random
motion that is produced by thermal motion. . . .
The fourth paper is only a rough draft right now,
and is about the electrodynamics of moving bodies
that employs a modified theory of space and
time.

> To Conrad Habicht, May 1905, giving him a
> foretaste of Einstein's *annus mirabilis*, during which
> he published, at the age of twenty-six, altogether
> five important papers that ushered physics into a
> new era. *CPAE*, Vol. 5, Doc. 27; see also Stachel
> et al., *Einstein's Miraculous Year*

From this we conclude that a balance-wheel clock
located at the Earth's equator must go more slowly,
by a very small amount, than a precisely similar
clock situated at one of the poles under otherwise
identical conditions.

> From "On the Electrodynamics of Moving
> Bodies." See *CPAE*, Vol. 2, Doc. 23 (p. 153 of
> translation volume). Originally in "Zur
> Elektrodynamik bewegter Körper," *Annalen der
> Physik* 17 (1905), 891–921. According to a letter I
> received from Professor Emeritus I. J. Good of

Virginia Tech in Blacksburg, Einstein neglected to
add that he was assuming the frame of reference of
the observer at the pole. In other inertial frames of
reference, the clock of the person on the equator
seems to go more slowly than that of the person
at the pole at least *some* of the time, but not
necessarily *all* of the time, as Einstein seems to be
saying. This lapse in exposition (or possibly a
mistake) led physicist Herbert Dingle astray; he
spent many years producing incorrect arguments
against the special theory of relativity.

Each ray of light moves in the coordinate system
"at rest" with the definite, constant velocity V independent of whether this ray of light is emitted by a
body at rest or a body in motion.

Ibid.; see *CPAE*, Vol. 2, Doc. 25 (p. 143 of
translation volume)

Thanks to my fortunate idea of introducing the relativity principle into physics, you (and others) now
enormously overrate my scientific abilities, to the
point where this makes me quite uncomfortable.

To Arnold Sommerfeld, January 14, 1908. *CPAE*,
Vol. 5, Doc. 73

* A physical theory can be satisfactory only if its
structures are composed of elementary foundations. The theory of relativity is ultimately as little
satisfactory as, for example, classical thermody-

namics was before Boltzmann had interpreted the entropy as probability.

Ibid.

People who have been privileged to contribute something to the advancement of science should not let [arguments about priority] becloud their joy over the fruits of common endeavor.

To Johannes Stark, February 22, 1908. A few days earlier, Einstein had expressed some annoyance that Stark failed to recognize Einstein's priority in regard to the relativistic relationship between mass and energy, which Stark had attributed to Max Planck in a paper in the *Physikalische Zeitschrift* in December 1907. See *CPAE*, Vol. 5, Doc. 88, and Doc. 70, n. 3

* It seems that scientific distinction and personal qualities do not always go hand in hand. I value a harmonious person far more than the craftiest formula jockey or experimentalist.

To Jakob Laub, March 16, 1910, lauding Laub's boss, Alfred Kleiner. *CPAE*, Vol. 5, Doc. 199

The more success the quantum theory has, the sillier it looks.

To Heinrich Zangger, May 20, 1912, reflecting Einstein's lack of faith in the quantum theory. *CPAE*, Vol. 5, Doc. 398

*The "theory of relativity" is correct insofar as the two principles upon which it is based are correct. Since these do seem to be largely correct, the theory of relativity in its present form seems to represent an important advance. I do not think that it has hampered the further development of theoretical physics!

> From "Reply to Comment by M. Abraham,"
> August 1912. *CPAE*, Vol. 4, Doc. 8

I cannot find the time to write because I am occupied with truly great things. Day and night I rack my brain in an effort to penetrate more deeply into the things that I gradually discovered in the past two years and that represent an unprecedented advance in the fundamental problems of physics.

> To Elsa Löwenthal, February 1914, about his work
> on an extension of his theory of gravitation, the
> first stage of which was published half a year
> earlier. *CPAE*, Vol. 5, Doc. 509

*Nature is showing us only the tail of the lion, but I have no doubt that the lion belongs to it even though, because of its large size, it cannot totally reveal itself all at once. We can see it only the way a louse that is sitting on it would.

> To Heinrich Zangger, March 10, 1914, regarding
> his work on the general theory of relativity. *CPAE*,
> Vol. 5, Doc. 513

* The principle of relativity can generally be phrased as: The laws of nature perceived by an observer are *independent* of his state of motion. . . . By combining the principle of relativity with the results of the constancy of light in a vacuum, one arrives by a purely deductive manner at what is called today "relativity theory." . . . Its significance lies in the fact that it provides conditions that every general law of nature must satisfy, for the theory teaches that natural phenomena are such that the laws do not depend on the state of motion of the observer to whom the phenomena are spatially and temporally related.

> *Vossische Zeitung*, April 26, 1914. *CPAE*, Vol. 6, Doc. 1

* One should not pursue goals that are easily achieved. One must develop an instinct for what one can just barely achieve through one's greatest efforts.

> To former student Walter Dällenbach, May 31, 1915, while giving him some advice on an electrical engineering project. *CPAE*, Vol. 8, Doc. 87

* Professionally, scientists and mathematicians are strictly international-minded and guard carefully against any unfriendly measures taken against their colleagues living in hostile foreign countries.

Historians and philologists, on the other hand, are chauvinistic hotheads.

To H. A. Lorentz, August 2, 1915, on the atmosphere in Berlin. *CPAE*, Vol. 8, Doc. 103

* In my personal experience I have hardly come to know the wretchedness of mankind better than as a result of this theory and everything connected to it. But it doesn't bother me.

To Heinrich Zangger, November 26, 1915, regarding the reception of the general theory of relativity. *CPAE*, Vol. 8, Doc. 152

* The theory is beautiful beyond comparison. However, only *one* colleague has really been able to understand it and [use it].

Ibid. The colleague was David Hilbert.

* Hardly anyone who truly understands it will be able to escape the charm of this theory.

From "Field Equations of Gravitation," November 1915, a paper further confirming the general theory of relativity by applying Riemann's curvature tensor

* Be sure you take a good look at them; they are the most valuable discovery of my life.

To Arnold Sommerfeld, December 19, 1915, regarding the equations in the above paper. *CPAE*, Vol. 8, Doc. 161

* The mainspring of scientific thought is not an external goal toward which one must strive, but the pleasure of thinking.

To Heinrich Zangger, ca. August 11, 1918. *CPAE*, Vol. 8, Doc. 597

* For me, a hypothesis is a statement whose *truth* is temporarily assumed, but whose *meaning* must be beyond all doubt.

To Edward Study, September 25, 1918. *CPAE*, Vol. 8, Doc. 624

The state of mind which enables a man to do work of this kind . . . is akin to that of the religious worshiper or the lover; the daily effort comes from no deliberate intention or program, but straight from the heart.

From "Principles of Research," a speech delivered at Max Planck's sixtieth birthday celebration, 1918. Published in *Mein Weltbild*, 109; reprinted in *Ideas and Opinions*, 227

I believe with Schopenhauer that one of the strongest motives that leads men to art and science is to escape from the rawness and monotony of every-day life and take refuge in a world crowded with the images of our own creation. . . . A finely tempered nature longs to escape from personal life into the world of objective perception and thought.

Ibid. Also in *Cosmic Religion*, 99

* [A researcher] adapts to the facts by intuitive selection of the possible theories based upon axioms.

"Induction and Deduction in Physics," *Berliner Tageblatt*, December 25, 1919. See also *CPAE*, Vol. 7, forthcoming

* The simplest picture one can form about the creation of an empirical science is along the lines of an inductive method. Individual facts are selected and grouped together such that their lawful connection becomes clearly apparent. By grouping these laws together, one can achieve other more general laws until a more or less uniform system for the available individual facts has been established. . . . However . . . the big advances in scientific knowledge originated this way only to a small degree. For, if a researcher were to approach things without a preconceived opinion, how would he be able to pick the facts from the tremendous richness of the most complicated experiences that are simple enough to reveal their connections through laws?

Ibid.

* The truly great advances in our understanding of nature originated in a way almost diametrically opposed to induction. The intuitive grasp of the essentials of a large complex of facts leads the scientist to the postulation of a hypothetical basic law, or several such laws. From these laws, he derives his conclusions, . . . which can then be compared to experience. Basic laws (axioms) and conclusions together form what is called a "theory." Every expert knows that the greatest advances in natural science . . . originated in this manner, and that their basis has this hypothetical character.

    Ibid.

* The *truth* of a theory can never be proven, for one never knows if future experience will contradict its conclusions.

    Ibid.

* When two theories are available and both are compatible with the given arsenal of facts, then there are no other criteria to prefer one over the other except the intuition of the researcher. Therefore one can understand why intelligent scientists, cognizant both of theories and of facts, can still be passionate adherents of opposing theories.

    Ibid.

Then I would feel sorry for the good Lord. The theory is correct anyway.

> In answer to the question of doctoral student Ilse
> Rosenthal-Schneider, in 1919, about how he would
> have reacted if his general theory of relativity had
> not been confirmed experimentally that year by
> Arthur Eddington and Frank Dyson. Quoted in
> Rosenthal-Schneider, *Reality and Scientific Truth*, 74

* [Constructive theories], from a relatively simple fundamental formalism, attempt to explain the more complex phenomena. . . . [Theories of principle, on the other hand,] are based on empirically discovered general properties of natural processes, on principles from which mathematically formulated criteria follow and that individual processes or their theoretical models must observe.

> Einstein's formulation of two kinds of scientific
> theories, 1919; he regarded constructive theories as
> more important, though each had its advantages.
> See "Was ist Relativitätstheorie?" in *Mein Weltbild*,
> 127, 128

* At present every coachman and every waiter argues about whether or not the relativity theory is correct.

> To Marcel Grossmann, September 12, 1920,
> expressing his surprise at the widespread interest
> in the general theory of relativity, which had
> caught the world's imagination. Because his theory
> was not understood, Einstein became an even

more mysterious figure. He continued to refer to
the public spectacle as the "relativity circus."
Einstein Archive 11-500

It is my inner conviction that the development of
science seeks in the main to satisfy the longing for
pure knowledge.

1920. In Moszkowski, *Conversations with
Einstein*, 173

The word "discovery" in itself is regrettable. For
discovery is equivalent to becoming aware of a
thing which is already formed; this links up with
proof, which no longer bears the character of "discovery" but, in the final analysis, of the means that
leads to discovery. . . . Discovery is really not a creative act.

Ibid., 95

The aspect of knowledge that has not yet been laid
bare gives the investigator a feeling akin to that experienced by a child who seeks to grasp the mas-
terly way in which adults manipulate things.

Ibid., 46

* The theory of relativity is nothing but another step
in the centuries-old evolution of our science, one
which preserves the relationships discovered in

the past, deepening their insights and adding new ones.

> In "Die hauptsächlichen Gedanken der Relativitätstheorie," an unpublished manuscript, ca. 1920, in the Einstein Archive

* As far as the laws of mathematics refer to reality, they are not certain; and as far as they are certain, they do not refer to reality.

> From "Geometry and Experience," an address to the Prussian Academy of Sciences, Berlin, January 27, 1921. In Einstein, *Sidelights on Relativity* (New York: Dover, 1983), 28. (In the last edition, in place of "the laws of mathematics," I used Philipp Frank's word "geometry," misquoted in *Einstein: His Life and Times*, 177. Thanks to a Czech reader for the correction.)

* We may in fact regard [geometry] as the most ancient branch of physics. . . . Without it I would have been unable to formulate the theory of relativity.

> Ibid., 32–33

* The four men who laid the foundations of physics on which I have been able to construct my theory are Galileo, Newton, Maxwell, and Lorentz.

> *New York Times*, April 4, 1921

The Lord God is subtle, but malicious he is not.

> Originally said to Princeton University
> mathematics professor Oscar Veblen, May 1921,
> while Einstein was in Princeton for a series of
> lectures, upon hearing that an experimental result
> by Dayton C. Miller of Cleveland, if true, would
> contradict his theory of gravitation. But the result
> turned out to be false. Some say by this remark
> Einstein meant that Nature hides her secrets by
> being subtle, while others say he meant that
> Nature is mischievous but not bent on trickery.
> Permanently inscribed in stone above the fireplace
> in the faculty lounge, 202 Jones Hall (called Fine
> Hall until Princeton's new mathematics building
> with the same name was constructed), in the
> original German: "Raffiniert ist der Herr Gott, aber
> boshaft ist Er nicht." Quoted widely in various
> translated versions, e.g., in Pais, *Subtle Is the Lord*;
> Frank, *Einstein: His Life and Times*, 285; and
> Hoffmann, *Albert Einstein: Creator and Rebel*, 146

I have second thoughts. Maybe God *is* malicious.

> To Valentine Bargmann, meaning that God makes
> us believe we have understood something that in
> reality we are far from understanding. Quoted in
> Sayen, *Einstein in America*, 51

Nature conceals her secrets because she is sublime,
not because she is a trickster.

> Aphorism scribbled in German in Einstein's hand;
> newly found (unnumbered) in the duplicate
> Einstein archive in Boston by József Illy. Later
> found in a letter to Oscar Veblen, April 30,

1930. Einstein Archive 23-153. Also translated as
"Nature conceals her secrets by exaltedness, but
not by cunning" in Fölsing, *Albert Einstein*, 503

Relativity is a purely scientific matter and has nothing to do with religion.

In response to the question of Randall Thomas
Davidson, the Archbishop of Canterbury, about
"what effect relativity would have on religion,"
London, 1921. Quoted in Frank, *Einstein: His Life
and Times*, 190. Armin Hermann, in *Albert Einstein*,
269, quotes more of the response: "Relativity
theory is an abstract science. It fits into every
worldview."

Now to the term "relativity theory." I admit that it is unfortunate, and has given occasion to philosophical misunderstandings.

To E. Zschimmer, September 30, 1921, referring to
Max Planck's term for his theory, which stuck
despite his unhappiness with it. He would have
preferred "theory of invariants," which he felt
better described the *method*, if not the content.
See Holton, *The Advancement of Science*, 69, 110,
312 n. 21

I was sitting in the patent office in Bern when all of a sudden a thought occurred to me: if a person falls freely, he won't feel his own weight. I was startled. This simple thought made a deep impression on me. It impelled me toward a theory of gravitation.

In Kyoto lecture, December 14, 1922. Translated
into English by Y. A. Ono in *Physics Today*, August
1932, from notes taken by Yon Ishiwara

* Describing the physical laws without reference to
geometry is similar to describing our thoughts
without words.

    Ibid.

* Because of the universal character of their subject
matter and their need for internationally organized
cooperation, [scientists] are inclined toward international understanding and therefore favor pacifist
goals.

    In Kurt Lenz and Walter Fabian, eds., *Die
    Friedensbewegung* (Berlin, 1922), 78–79

* Technology resulting from the sciences has internationally chained together economies, and this has
caused all wars to become a matter of international
importance. When this situation has entered the
consciousness of mankind, after sufficient turmoil,
then men will also find the energy and goodwill to
create organizations that have the power to end
wars.

    Ibid.

* The theory of relativity states: The laws of nature are to be formulated free of any specific coordinates because a coordinate system does not conform to anything real. The simplicity of a hypothetical law is to be judged only according to its generally covariant form. . . . The laws of nature have never had and still do not have a preferential coordinate system. . . . The theory of relativity claims only that the *general* laws of nature are the same with respect to any system.

> *Annalen der Physik* 69 (1922), 438

* There is always a certain charm in tracing the evolution of theories in the original papers; often such study offers deeper insights into the subject matter than the systematic presentation of the final result, polished by the words of many contemporaries.

> Foreword to the Japanese edition of Einstein's papers, published May 1923

* In seeking an integrated theory, the intellect cannot rest contentedly with the assumption that there are two distinct fields, totally independent of each other by their nature.

> From his delayed Nobel Lecture, written June 11, 1923, and given July 1923 in Göteborg. This statement foretold Einstein's lifelong search for a unified field theory of gravity and electromagnetism. See *Les Prix Nobel en 1921–1922* (Stockholm, 1923)

After a certain high level of technical skill is achieved, science and art tend to coalesce in esthetics, plasticity, and form. The greatest scientists are artists as well.

Remark made in 1923. Recalled by Archibald Henderson, *Durham Morning Herald*, August 21, 1955. Einstein Archive 33-257

The more one chases after quanta, the better they hide themselves.

To Paul Ehrenfest, July 12, 1924, expressing his frustration over quantum theory. Einstein Archive 10-089

My interest in science was always essentially limited to the study of principles. . . . That I have published so little is due to this same circumstance, as the great need to grasp principles has caused me to spend most of my time on fruitless pursuits.

To Maurice Solovine, October 30, 1924. Einstein Archive 21-195; published in *Letters to Solovine*, 63

Quantum mechanics is very worthy of regard. But an inner voice tells me that this is not yet the right track. The theory yields much, but it hardly brings us closer to the Old One's secrets. I, in any case, am convinced that *He* does not play dice.

To Max Born, December 4, 1926. In Born, *Born-Einstein Letters*, 91

I admire to the highest degree the achievement of the younger generation of physicists which goes by the term quantum mechanics, and believe in the deep level of truth of that theory; but I believe that its limitation to *statistical laws* will be a temporary one.

From a speech on June 28, 1929, on acceptance of the Planck Medal. Quoted in *Forschungen und Fortschritte* 5 (1929), 248–249

* The main source of all technological achievements is the divine curiosity and playful drive of the tinkering and thoughtful researcher, as much as it is the creative imagination of the inventor.

August 22, 1930, in a radio broadcast in Berlin. Transcribed by Friedrich Herneck in *Die Naturwissenschaften* 48 (1930), 33

* Those who thoughtlessly make use of the miracles of science and technology, without understanding more about them than a cow eating plants understands about botany, should be ashamed of themselves.

Ibid.

Concern for man himself and his fate must always constitute the chief objective of all technological endeavors ... in order that the creations of our mind shall be a blessing and not a curse to mankind.

Never forget this in the midst of your diagrams and equations.

> From an address entitled "Science and Happiness,"
> presented at the California Institute of Technology,
> Pasadena, February 16, 1931. Quoted in the *New
> York Times*, February 17 and 22, 1931

Why does this magnificent applied science which saves work and makes life easier bring us so little happiness? The simple answer: because we have not yet learned to make sensible use of it.

> In reference to technology. Ibid.

* There is no doubt that all but the crudest scientific work is based on a firm belief—akin to a religious feeling—in the rationality and comprehensibility of the world.

> "On Science," in *Cosmic Religion* (1931), 98

The scientist finds his reward in what Henri Poincaré calls the joy of comprehension, and not in the possibilities of application to which any discovery may lead.

> From the epilogue to Planck, *Where Is Science
> Going?* (1932), 211

I believe that the present fashion of applying the methods of physics to human life is not only a mistake but heinous.

> On a "worldview" of relativity and the gross abuse
> of physical science in areas in which it is not
> applicable. Ibid.; also quoted by Loren Graham in
> Holton and Elkana, *Albert Einstein: Historical and
> Cultural Perspectives*, 107

* The creative principle [of science] resides in mathematics.

> From the Herbert Spencer Lecture, Oxford,
> June 10, 1933. Published in *Ideas and Opinions*, 274

The years of anxious searching in the dark for a truth that one feels but cannot express, the intense desire and the alternations of confidence and misgiving until one achieves clarity and understanding, can be understood only by those who have experienced them.

> From a lecture at the University of Glasgow,
> June 20, 1933. Published in *The Origins of the Theory
> of Relativity*; reprinted in *Mein Weltbild*, 138, and in
> *Ideas and Opinions*, 289–290

* It is not the *result* of scientific research that ennobles humans and enriches their nature, but the *struggle to understand* while performing creative and open-minded intellectual work.

> From "Good and Evil." Published in *Mein Weltbild*
> (1934), 14; reprinted in *Ideas and Opinions*, 12

The general public may be able to follow the details of scientific research to only a modest degree; but it can register at least one great and important notion: the confidence that human thought is dependable and natural law is universal.

From "Science and Society," 1935. Reprinted in
Einstein on Humanism, 13

* Scientific research is based on the assumption that all events, including the actions of mankind, are determined by the laws of nature.

To student Phyllis Wright, January 24, 1936.
Einstein Archive 52-337

All of science is nothing more than the refinement of everyday thinking.

From "Physics and Reality," *Journal of the Franklin Institute* 221, no. 3 (March 1936), 349–382

* It is always delightful when a great and beautiful idea proves to be consonant with reality.

To Sigmund Freud, April 21, 1936. Einstein
Archive 32-566

* We (Mr. Rosen and I) sent our publication to you without the authorization that you may show it to other specialists before it is printed. I do not see any reason to follow your anonymous reviewer's recommendations (which incidentally are erroneous).

In view of the foregoing, I will consider having the work published elsewhere.

> To the editor of the *Physical Review*, July 27, 1936.
> The article, "On Gravitational Waves," with
> Nathan Rosen, was later published in the *Journal
> of the Franklin Institute* 223 (1937), 43–54. Einstein
> Archive 19-087

I still struggle with the same problems as ten years ago. I succeed in small matters but the real goal remains unattainable, even though it sometimes seems palpably close. It is hard yet rewarding: hard because the goal is beyond my abilities, but rewarding because it makes one oblivious to the distractions of everyday life.

> To Otto Juliusburger, September 28, 1937. Einstein
> Archive 38-163

Physical concepts are free creations of the human mind and are not, however it may seem, uniquely determined by the external world.

> From *The Evolution of Physics*, with Leopold Infeld
> (1938)

What we call physics comprises that group of natural sciences which base their concepts on measurements, and whose concepts and propositions lend themselves to mathematical formulations.

> From "The Fundamentals of Theoretical Physics,"
> *Science* 91 (May 24, 1940), 487–492

There has always been an attempt to find a unifying theoretical basis for all these [various branches of physics] . . . from which all the concepts and relationships among the individual disciplines might be derived by a logical process. This is what we mean by the search for a foundation of the whole of physics. The confident belief that this ultimate goal may be reached is the wellspring of the passionate devotion that has always motivated the researcher.

Ibid.

You cannot love a car the way you love a horse. The horse, unlike a machine, compels human emotions. A machine disregards human feelings. . . . Machines make our life impersonal, stunt certain qualities in us, and create an impersonal environment.

From a conversation recorded by Algernon Black,
Fall 1940. Einstein Archive 54-834

It is hard to sneak a look at God's cards. But that he would choose to play dice with the world . . . is something I cannot believe for a single moment.

To Cornel Lanczos, March 21, 1942, expressing his
reaction to quantum theory, which refutes
relativity theory by stating that an observer *can*
influence reality, that events *do* happen randomly.
Einstein Archive 15-294. Quoted in Hoffmann,
Albert Einstein: Creator and Rebel, chapter 10; Frank,
Einstein: His Life and Times, 208, 285; and Pais,
Einstein Lived Here, 114. My favorite variant of this
quotation, sent to me by a rabbi, is: "God doesn't

play craps with the universe." Physicist Niels Bohr
is said to have told Einstein, "Stop telling God
what to do!"

* I never understood why the theory of relativity,
with its concepts and problems so far removed
from practical life, should have met with such a
lively, indeed passionate, reception among a broad
segment of the public.

Written in 1942. Published in preface to Frank,
Einstein: His Life and Times, 1979 edition

Do not worry about your difficulties in mathematics; I can assure you that mine are still greater.

To junior high school student Barbara Wilson,
January 7, 1943. Einstein Archive 42-606; also
quoted in Dukas and Hoffmann, *Albert Einstein, the
Human Side*, 8

The entire history of physics since Galileo bears
witness to the importance of the function of the theoretical physicist, from whom the basic theoretical
ideas originate. A priori construction in physics is
as essential as empirical facts.

Memo written with Hermann Weyl to the faculty
of the Institute for Advanced Study, early 1945,
recommending theoretician Wolfgang Pauli over
Robert Oppenheimer for a professorship at the
Institute. Pauli declined, and Oppenheimer, who

was offered the directorship in 1946, accepted.
Quoted in Regis, *Who Got Einstein's Office?* 135

* The theory of relativity, as I developed it originally, still does not explain atomism and the quantum phenomena. And neither does it include a common mathematical formulation covering the phenomena of both the electromagnetic and gravitational fields. This demonstrates that the original formulation of the theory of relativity is not definitive . . . its means of expression are in process of evolution. . . . The task to which I am now giving my greatest efforts is to resolve the dualism between the theories of gravitation and electromagnetism, and to reduce them to the one and same mathematical form.

> From an interview with Alfred Stern, *Contemporary Jewish Record* 8 (June 1945), 245–249

* I am not a positivist. Positivism states that what cannot be observed does not exist. This conception is scientifically indefensible, for it is impossible to make valid affirmations of what people "can" or "cannot" observe. One would have to say "only what we observe exists," which is obviously false.

> Ibid.

* I sold myself body and soul to Science—a flight from the "I" and "we" to the "it."

> To Hermann Broch, September 2, 1945. Einstein
> Archive 34-048.1; also quoted in Hoffmann, *Albert*
> *Einstein: Creator and Rebel*, 254

A scientific person will never understand why he should believe opinions only because they are written in a certain book. [Furthermore], he will never believe that the results of his own attempts are final.

> To J. Lee, September 10, 1945. Einstein Archive
> 57-061

* As a scientist, I believe that nature is a perfect structure, seen from the standpoint of reason and logical analysis.

> To Raymond Benenson, January 31, 1946. Einstein
> Archive 56-505

I believe that the abominable deterioration of ethical standards stems primarily from the mechanization and depersonalization of our lives—a disastrous by-product of science and technology. Nostra culpa!

> To Otto Juliusburger, April 11, 1946. Einstein
> Archive 38-228

Science will stagnate if it is made to serve practical goals.

In answer to a question posed by the Overseas News Agency, January 20, 1947. Quoted in Nathan and Norden, *Einstein on Peace*, 402

* If God had been satisfied with inertial systems, he would not have created gravitation.

Said to Abraham Pais, 1947. See Pais, *A Tale of Two Continents*, 227

* I believe that this is the God-given generalization of general relativity theory. Unfortunately, the Devil comes into play, since one cannot solve the [new] equations.

On his most recent efforts to generalize general relativity to a so-called unified field theory. Ibid.

* I do not like it when it can be done this way or that way. It should be: This way or not at all.

On theories in general. Ibid.

In my scientific work, I am still hampered by the same mathematical difficulties that have been making it impossible for me to confirm or refute my general relativistic field theory.... I won't ever

solve it; it will be forgotten and must later be rediscovered again.

To Maurice Solovine, November 25, 1948. Einstein
Archive 21-256, 80-865; published in *Letters to
Solovine*, 105, 107

* The supreme task of the physicist is to arrive at
those universal elementary laws from which the
cosmos can be built up by pure deduction.

From an interview with Alfred Werner, *Liberal
Judaism* 16 (April–May 1949), 12

The grand aim of all science is to cover the greatest
number of empirical facts by logical deduction
from the smallest number of hypotheses or axioms.

Quoted in *Life* magazine, January 9, 1950

The unified field theory has been put into retirement. It is so difficult to employ mathematically
that I have not been able to verify it somehow, in
spite of all my efforts. This state of affairs will no
doubt last many more years, mostly because physicists have little understanding of logical-philosoph-
ical arguments.

To Maurice Solovine, February 12, 1951. Einstein
Archive 21-277; published in *Letters to Solovine*, 123

Science is a wonderful thing if one does not have to earn a living at it. One should earn one's living by work of which one is sure one is capable. Only when we do not have to be accountable to anyone can we find joy in scientific endeavor.

To a California student, 1951. Quoted in Dukas and Hoffmann, *Albert Einstein, the Human Side*, 57

The betterment of conditions the world over is not strictly dependent on scientific knowledge but on the fulfillment of human traditions and ideals.

1952. Quoted in French, *Einstein: A Centenary Volume*, 197

Development of Western science is based on two great achievements: the invention of the formal logical system (in Euclidean geometry) by the Greek philosophers, and the discovery of the possibility of finding out causal relationships by systematic experiment (during the Renaissance).

To J. S. Switzer, April 23, 1953. Einstein Archive 61-381

That no one can make a definite statement about [the unified field theory's] confirmation or nonconfirmation results from the fact that there are no methods of affirming anything with respect to solutions that do not yield to the peculiarities of such a complicated nonlinear system of equations. It is even possible that no one will ever know.

To Maurice Solovine, May 28, 1953. Einstein
Archive 21-300; published in *Letters to Solovine*, 149

In striving to do scientific work, the chance—even for very gifted persons—to achieve something of real value is very small. . . . There is only one way out: devote most of your time to some practical work . . . that agrees with your nature, and spend the rest of it in study. So you will be able . . . to lead a normal and harmonious life even without the special blessings of the Muses.

To a man in India who was unsure about what
lifework to pursue, July 14, 1953. Quoted in Dukas
and Hoffmann, *Albert Einstein, the Human Side*, 59

It is strange that science, which in the old days seemed harmless, should have evolved into a nightmare that causes everyone to tremble.

To Queen Elizabeth of Belgium, March 28, 1954.
Einstein Archive 32-410; quoted in Whitrow,
Einstein, 89

I believe that every true theorist is a kind of tamed metaphysicist, no matter how pure a "positivist" he may fancy himself to be.

From "On the Generalized Theory of Gravitation,"
Scientific American 182, no. 4 (April 1954)

* There is no doubt that the special theory of relativity, if we look at its development in retrospect, was ripe for discovery in 1905.

> To Carl Seelig, February 19, 1955. Einstein Archive 39-069

It appears doubtful that a [classical] field theory can account for the atomistic structure of matter and radiation as well as of quantum phenomena. Most physicists will reply with a firm "no," since they believe that the quantum problem has been solved in principle by other means. However that may be, Lessing's comforting words stay with us: "The struggle for truth is more precious than its assured possession."

> Einstein's final written scientific words, on quantum theory, March 1955, about a month before his death. Published in Seelig, *Helle Zeit, dunkle Zeit*; also quoted by Pais in French, *Einstein: A Centenary Volume*, 37

* When I am judging a theory, I ask myself whether, if I were God, I would have arranged the world in such a way.

> Said to his assistant Banesh Hoffmann. See Harry Woolf, ed., *Some Strangeness in the Proportion* (Reading, Mass.: Addison-Wesley, 1980), 476

Men really devoted to the progress of knowledge concerning the physical world . . . never worked for practical, let alone military, goals.

> Ibid., 510

* When we say that we understand a group of natural phenomena, we mean that we have found a constructive theory that embraces them.

> "What Is the Theory of Relativity?" in *Out of My Later Years*, 54–55

I have thought a hundred times as much about the quantum problems as I have about general relativity theory.

> To Otto Stern. Quoted by Pais in French, *Einstein: A Centenary Volume*, 37

I can, if the worse comes to worst, still realize that God may have created a world in which there are no natural laws. In short, chaos. But that there should be statistical laws with definite solutions, i.e., laws that compel God to throw dice in each individual case, I find highly disagreeable.

> To James Franck. Quoted by C. P. Snow in French, *Einstein: A Centenary Volume*, 6

* All physical theories, their mathematical expressions notwithstanding, ought to lend themselves to so simple a description that even a child could understand them.

> Attributed to Einstein by Louis de Broglie in
> *Nouvelles perspectives en microphysique* (trans. New
> York: Basic Books, 1962), 184. Also in Clark,
> *Einstein*, 344

One thing I have learned in a long life: that all our science, measured against reality, is primitive and childlike—and yet it is the most precious thing we have.

> Quoted in Hoffmann, *Albert Einstein: Creator and
> Rebel*, v

It follows from the theory of relativity that mass and energy are both different manifestations of the same thing—a somewhat unfamiliar conception for the average man. Furthermore, $E = mc^2$, in which energy is put equal to mass multiplied with the square of the velocity of light, showed that a very small amount of mass may be converted into a very large amount of energy ... the mass and energy in fact were equivalent.

> Read aloud to an audience, filmed and shown in
> *Nova*'s Einstein biography, 1979

Physics is essentially an intuitive and concrete science. Mathematics is only a means for expressing the laws that govern phenomena.

Quoted by Maurice Solovine in "Introduction" to *Letters to Solovine*, 7–8

In the beginning (if there was such a thing), God created Newton's laws of motion together with the necessary masses and forces. This is all; everything beyond this follows from the development of appropriate mathematical methods by means of deduction.

From "Autobiographical Notes," in Schilpp, *Albert Einstein: Philosopher-Scientist*, 19

A theory is the more impressive the greater the simplicity of its premises, the more different kinds of things it relates, and the more extended its area of applicability.

Ibid., 33. Einstein often refers to the value of simple hypotheses, believing they may become basic traits of future theoretical representations, as in the case of the emission and absorption of radiation. See *CPAE*, Vol. 6, Doc. 34; see also what may be a paraphrase of this general idea, the quotation on simplicity in "Attributed to Einstein" at the back of the book.

* [Classical thermodynamics] is the only physical theory of universal content that I am convinced, within the framework of its basic concepts, will never be overthrown.

> Ibid.

An hour sitting with a pretty girl on a park bench passes like a minute, but a minute sitting on a hot stove seems like an hour.

> Einstein's explanation of relativity that he gave to his secretary, Helen Dukas, to relay to reporters and other laypersons. Quoted in Sayen, *Einstein in America*, 130

The aim of science is, on the one hand, as complete a comprehension as possible of the connection between perceptible experiences in their totality, and, on the other hand, the achievement of this aim by employing a minimum of primary concepts and relations.

> In *Ideas and Opinions*, 293–294

I have little patience for scientists who take a board of wood, look for its thinnest part, and drill a great number of holes when the drilling is easy.

> Related by Philipp Frank in "Einstein's Philosophy of Science," *Reviews of Modern Physics* (1949)

* It would be possible to describe everything scientifically, but it would make no sense. It would be a description without meaning—as if you described a Beethoven symphony as a variation of wave pressure.

Quoted in Max Born, *Physik im Wandel meiner Zeit* (Braunschweig: Vieweg, 1966)

A scientist is a mimosa when he himself has made a mistake, and a roaring lion when he discovers a mistake of others.

Quoted in Ehlers, *Liebes Hertz!* 45

Einstein partaking in his favorite hobby, sailing, in 1936. (Courtesy of AIP Emilio Segrè Visual Archives)

ABORTION

A woman should be able to choose to have an abortion up to a certain point in pregnancy.

To the World League for Sexual Reform, Berlin, September 6, 1929. Einstein Archive 48-304; also quoted in Grüning, *Ein Haus für Albert Einstein*, 305

ACHIEVEMENT

The value of achievement lies in the act of achieving.

October 1950. Einstein Archive 60-297

AGING

*I have remained a simple fellow who asks nothing of the world; only my youth is gone—the enchanting youth that forever walks on air.

To Anna Meyer-Schmid, May 12, 1909. *CPAE*, Vol. 5, Doc. 154

There is, after all, something eternal that lies beyond the reach of the hand of fate and of all human delusions. And such eternals lie closer to an older

person than to a younger one who oscillates between fear and hope.

> To Queen Elizabeth of Belgium, March 20, 1936.
> Einstein Archive 32-387; also quoted in *Einstein: A
> Portrait*, 54

People like you and I, though mortal of course, like
everyone else, do not grow old no matter how long
we live. What I mean is that we never cease to stand
like curious children before the great Mystery into
which we were born.

> To Otto Juliusburger, September 29, 1942. Einstein
> Archive 38-238

I am content in my later years. I have kept my good
humor and take neither myself nor the next person
seriously.

> To P. Moos, March 30, 1950. Einstein Archive
> 60-587

I have always loved solitude, a trait that tends to
increase with age.

> To E. Marangoni, October 1, 1952. Einstein Archive
> 60-406

If younger people were not taking care of me, I
would surely try to be institutionalized, so that I
would not have to become so concerned about the

decline of my physical and mental powers, which after all is unpreventable in the natural course of things.

> To W. Lebach, May 12, 1953. Einstein Archive 60-221

In one's youth every person and every event appear to be unique. With age, one becomes much more aware that similar events recur. Later on, one is less often delighted or surprised, but also less disappointed.

> To Queen Elizabeth of Belgium, January 3, 1954. Einstein Archive 32-408

I believe that older people who have scarcely anything to lose ought to be willing to speak out on behalf of those who are young and who are subject to much greater restraint.

> To Queen Elizabeth of Belgium, March 28, 1954. Einstein Archive 32-411

Even [old] age has very beautiful moments.

> To Margot Einstein. Quoted in Sayen, *Einstein in America*, 298

I live in that solitude which is painful in youth, but delicious in maturity.

> Quoted in *Out of My Later Years*, 13

AMBITION

Nothing truly valuable arises from ambition or from a mere sense of duty; it stems rather from love and devotion toward men and toward objective things.

> Statement for an Idaho farmer who requested some words that his son, Albert Wada, could live by as he grew up, July 30, 1947. Einstein Archive 58-934; quoted in Dukas and Hoffmann, *Albert Einstein, the Human Side*, 46

ANIMALS/PETS

Thank you very much for your kind and interesting information. I am sending my heartiest greetings to my namesake, also from our tomcat, who was very interested in the story and even a little jealous. The reason is that his own name, "Tiger," does not express, as in your case, the close kinship to the Einstein family.

> To Edward Moses, August 10, 1946, after learning that his ship's crew had rescued a kitten in Germany and named it Einstein. Einstein Archive 57-194

I know what's wrong, dear fellow, but I don't know how to turn it off.

To his tomcat, Tiger, who seemed depressed
because he was housebound by rain. Recalled by
Ernst Straus in his memorial talk, "Albert Einstein,
the Man," at UCLA, May 1955, 14–15

The main thing is that *he* knows.

About a friend's dog, Moses, whose long fur made
it difficult to tell one end from the other. From an
interview, January 15, 1979, with Margot Einstein
by J. Sayen; quoted in Sayen, *Einstein in America*,
131

The dog is very smart. He feels sorry for me because I receive so much mail; that's why he tries to
bite the mailman.

Regarding his dog, Chico. Quoted in Ehlers, *Liebes
Hertz!* 162

ART AND SCIENCE

* If what is seen and experienced is portrayed in the
language of logic, then it is science. If it is communicated through forms whose connections are not
accessible to the conscious mind but are recognized
intuitively, then it is art.

To the editor of the German magazine *Menschen*
4 (January 1921). *CPAE*, Vol. 7, Doc. 49

ASTROLOGY

* The reader should note [Kepler's] remarks on astrology. They show that the inner enemy, conquered and rendered innocuous, was not yet completely dead.

> From the "Introduction" to *Johannes Kepler: Life and Letters* by Carola Baumgart (New York: Philosophical Library, 1951). See also "Attributed to Einstein" at the back of the book

AUTHORITY

* A foolish faith in authority is the worst enemy of truth.

> To Jost Winteler, July 8, 1901. *CPAE*, Vol. 1, Doc. 115

BIRTH CONTROL

I am convinced that some political and social activities and practices of the Catholic organizations are detrimental and even dangerous for the community as a whole, here and everywhere. I mention here only the fight against birth control at a time when overpopulation in various countries has become a serious threat to the health of people and a grave obstacle to any attempt to organize peace on this planet.

To a reader of the Brooklyn *Tablet*, the newspaper of the diocese of Brooklyn and Queens, 1954, who questioned Einstein about whether he had been correctly quoted on the subject

BIRTHDAYS

My dear little sweetheart ... first, my belated cordial congratulations on your birthday yesterday, which I had forgotten once again.

To girlfriend Mileva Marić, his future wife, December 19, 1901. *CPAE*, Vol. 1, Doc. 130

* What is there to celebrate? Birthdays are automatic things. Anyway, birthdays are for children.

New York Times, March 12, 1944

* My birthday was a natural disaster, a shower of paper, full of platitudes, under which one almost drowned.

To Hans Muehsam, March 30, 1954. Einstein Archive 38-434

BLACKS/RACISM/SLAVERY

Insofar as we may at all claim that slavery has been abolished today, we owe its abolition to the practical consequences of science.

From "Science and Society," 1935. Quoted in
Einstein on Humanism, 11

This country still has a heavy debt to discharge for all the troubles and disabilities it has laid on the Negro's shoulder. . . . To the Negro and his wonderful songs and choirs we owe the finest contribution in the realm of art which America has given the world.

At the dedication of the Wall of Fame at the 1940
World's Fair

[Bias against the Negro] is the worst disease from which the society of our nation suffers.

Quoted in the *New York Times*, September 23, 1946

Security against lynching is one of the most urgent tasks of our generation.

To President Harry Truman, in a letter presented
to him by civil rights activist, athlete, and singer
Paul Robeson. Quoted in the *New York Times*,
September 23, 1946

BOOKS

What I have to say about this book can be found inside the book.

> Reply to a *New York Times* reporter's request for a comment on Einstein's book *The Evolution of Physics*, written with Leopold Infeld. Quoted in Ehlers, *Liebes Hertz!* 65

CAUSALITY

* The causal way of looking at things always answers only the question, "Why?" but never "To what end?" ... However, if someone asks, "For what purpose should we help one another, make life easier for each other, make beautiful music together, have inspired thoughts?" he would have to be told, "If you don't feel the reasons, no one can explain them to you." Without this primary feeling we are nothing and had better not live at all.

> To Hedwig Born, September 19, 1919. In Born, *Born-Einstein Letters*, 13

* I believe that whatever we do or live for has its causality. It is good, however, that we do not know what it is.

> In conversation with Indian mystic, poet, and musician Rabindranath Tagore. Published in *Asia* 31 (March 1931)

CLARITY

* All my life I have been a friend of well-chosen, sober words and of concise presentation. Pompous phrases and words give me goose bumps whether they deal with the theory of relativity or with anything else.

> *Berliner Tageblatt*, August 27, 1920, 1–2. See also
> *CPAE*, Vol. 7, forthcoming

CLOTHES

If I were to start taking care of my grooming, I would no longer be my own self.... So the hell with it. If you find me so repulsive, then look for a boyfriend who is more appealing to female tastes. But I will continue to be unconcerned about it, which surely has the advantage that I'm left in peace by many a fop who would otherwise come to see me.

> To future second wife, Elsa Löwenthal,
> ca. December 2, 1913. *CPAE*, Vol. 5, Doc. 489

* Only a certain regimen regarding attire, etc., so as not to be counted among the rejects of the local human race, disturbs my peace of mind somewhat.

> To the Hurwitz family, May 4, 1914, on his new life
> in Berlin. *CPAE*, Vol. 8, Doc. 6

I like neither new clothes nor new kinds of food.

> Quoted in Pais, *Subtle Is the Lord*, 16

It would be a sad situation if the wrapper were better than the meat wrapped inside it.

> Recalled in Einstein's obituary in the *New York Times*, April 19, 1955, about Einstein's notorious disregard for his outward appearance

"Why should I? Everyone knows me there" (upon being told by his wife to dress properly when going to the office). "Why should I? No one knows me there" (upon being told to dress properly for his first big conference).

> Quoted in Ehlers, *Liebes Hertz!* 87

I have reached an age when, if someone tells me to wear socks, I don't have to.

> Quoted by neighbor and fellow physicist Allan Shenstone, in Sayen, *Einstein in America*, 69

When I was young I found out that the big toe always ends up making a hole in a sock. So I stopped wearing socks.

> To Philippe Halsman. Quoted in French, *Einstein: A Centenary Volume*, 27

COMPETITION

I no longer need to take part in the competition of the big brains. Participating [in the process] has always seemed to me to be an awful type of slavery no less evil than the passion for money or power.

> To Paul Ehrenfest, May 5, 1927, regarding the rat race for academic promotions. Einstein Archive 10-163; also quoted in Dukas and Hoffmann, *Albert Einstein, the Human Side*, 60

COMPREHENSIBILITY

The eternal mystery of the world is its comprehensibility.... The fact that it is comprehensible is a miracle.

> From "Physics and Reality," *Journal of the Franklin Institute* 221, no. 3 (March 1936), 349–382. Reprinted in *Ideas and Opinions*, 292. Popularly paraphrased as, "The most incomprehensible thing about the universe is that it is comprehensible."

CONFORMITY

* The undignified mania of adaptive conformity, exhibited by many of my social class, has always been very repulsive to me.

Jüdische Rundschau, June 21, 1921. See also *CPAE*,
Vol. 7, forthcoming. It is odd to see Einstein
referring to himself as belonging to a "social class,"
since he generally abhorred such concepts. Most
likely it was one of his tongue-in-cheek
expressions. Others may not have been surprised
by the above allusion, however: some Soviet
intellectuals during the Stalin era mocked his
political ideals as "Bourgeois Einsteinism"; see
A. Heilbut, *Exiled in Paradise* (Boston: Beacon Press,
1983).

CONSCIENCE

Never do anything against conscience even if the
state demands it.

From a *Saturday Review* obituary of Einstein,
April 30, 1955

CREATIVITY

The monotony of a quiet life stimulates the creative
mind.

From a speech, "Civilization and Science," at
Royal Albert Hall, London, October 3, 1933.
Quoted in *The Times* (London), October 4, 1933, 14

Without creative personalities able to think and
judge independently, the upward development of

society is as unthinkable as the development of the
individual personality without the nourishing soil
of the community.

> Published in *Mein Weltbild* (1934), 12; reprinted in
> *Ideas and Opinions*, 14

True art is characterized by an irresistible urge in
the creative artist.

> November 15, 1950, regarding musician Ernst
> Bloch. Einstein Archive 34-332; also quoted in
> Dukas and Hoffmann, *Albert Einstein, the Human
> Side*, 77

CRIMINALS

I think we have to safeguard ourselves against people who are a menace to others, quite apart from
what may have motivated their deeds.

> To Otto Juliusburger, April 11, 1946. Einstein
> Archive 38-228

CURIOSITY

The important thing is not to stop questioning. Curiosity has its own reason for existing. One cannot
help but be in awe when one contemplates the mysteries of eternity, of life, of the marvelous structure

of reality. It is enough if one tries to comprehend only a little of this mystery every day.

From the memoirs of William Miller, an editor, quoted in *Life* magazine, May 2, 1955

Curiosity is a delicate little plant which, aside from stimulation, stands mainly in need of freedom.

Quoted in Cline, *Men Who Made a New Physics*, 64

DEATH PENALTY

I have reached the conviction that the abolition of the death penalty is desirable. Reasons: (1) Irreversibility in the event of an error in justice; (2) detrimental moral influence on those who . . . have to carry out the procedure.

To a Berlin publisher, November 3, 1927. Einstein Archive 46-009. However, several months earlier, according to the *New York Times*, the story had been a bit different: "Professor Einstein does not favor the abolition of the death penalty. . . . He could not see why society should not rid itself of individuals proved socially harmful. He added that society had no greater right to condemn a person to life imprisonment than it had to sentence him to death"; see the *New York Times*, March 6, 1927; also noted in Pais, *Einstein Lived Here*, 174

I am not for punishment at all, but only for measures that save society and protect it. In principle, I would not be opposed to killing individuals who are worthless or dangerous in that sense. I am against it only because I do not trust people, i.e., the courts. What I value in life is quality rather than quantity.

To Valentine Bulgakov, November 4, 1931.
Einstein Archive 45-702

THE ENGLISH AND THE
ENGLISH LANGUAGE

* Whereas in Germany, in general, the judgment of my theory depended upon the political orientation of the newspapers, the English scientists' attitude has proven that their sense of objectivity cannot be muddled by political viewpoints.

Jüdische Rundschau, June 21, 1921. See *CPAE*, Vol. 7,
forthcoming

I cannot write in English, because of the treacherous spelling. When I am reading, I only hear it and am unable to remember what the written word looks like.

To Max Born, September 7, 1944, expressing the
difficulty he had with the language of his adopted
country, even though he was eager to become an
American citizen. In Born, *Born-Einstein Letters*, 148

EPISTEMOLOGY

* When I think of the most able students I have encountered in my teaching—I mean those who have
distinguished themselves not only by skill but by
independence of thought—then I must confess that
all have had a lively interest in epistemology. No
one can deny that epistemologists have paved the
road for progress [toward the theory of relativity];
Hume and Mach, at least, have helped me considerably, both directly and indirectly.

From "Ernst Mach," *Physikalische Zeitschrift* 17
(1916)

Epistemology without contact with science becomes an empty scheme. Science without epistemology is—insofar as it is thinkable at all—primitive and muddled.

From "Autobiographical Notes," in Schilpp, *Albert
Einstein: Philosopher-Scientist*, 5

FLYING SAUCERS

Those people have seen *something*. What it is, I do
not know and I am not curious to know.

To L. Gardner, July 23, 1952. Einstein Archive
59-803. Einstein also believed that people should
not read science fiction—that it distorts science

and gives people the *illusion* of understanding
science. See letter to a boy in Iowa, Einstein
Archive, 59th reel

FORCE

Force always attracts men of low morality, and I believe it to be an invariable rule that tyrants of genius are succeeded by scoundrels.

From "What I Believe," *Forum and Century* 84
(1930), 193-194; reprinted in *Ideas and Opinions*,
8–11

GAMES

I do not play games. . . . There is no time for it.
When I get through with work, I don't want anything that requires the working of the mind.

New York Times, March 28, 1936

GOOD ACTS

Good acts are like good poems. One may easily
get their drift, but they are not always rationally
understood.

To Maurice Solovine, April 9, 1947. Einstein
Archive 21-250; published in *Letters to Solovine*, 99,
101

HOME

* It is not so important where one settles down. The best thing is to follow your instincts without too much reflection.

> To Max Born, March 3, 1920. In Born, *Born-Einstein Letters*, 26

HOMOSEXUALITY

Homosexuality should not be punishable except to protect children.

> To the World League for Sexual Reform, Berlin, September 6, 1929. Einstein Archive 48-304; also quoted in Grüning, *Ein Haus für Albert Einstein*, 305–306

INDIVIDUALS/INDIVIDUALITY

The really valuable thing in the pageant of human life seems to me not the political state, but the creative, sentient individual, the personality; it alone creates the noble and the sublime, while the herd as such remains dull in thought and dull in feeling.

> From "What I Believe," *Forum and Century* 84 (1930), 193–194; reprinted in *Ideas and Opinions*, 8–11

Valuable achievement can sprout from human society only when it is sufficiently loosened to make possible the free development of an individual's abilities.

From an unpublished article on tolerance, 1934.
Einstein Archive 49-094

While it is true that an inherently free ... person may be destroyed, such an individual can never be enslaved or used as a blind tool.

Statement in *Impact*, UNESCO, 1950

It is important for the common good to foster individuality: for only the individual can produce the new ideas which the community needs for its continuous improvement and requirements—indeed, to avoid sterility and petrification.

From a message for a Ben Schemen dinner, March 1952. Einstein Archive 28-932

INTELLIGENCE

It is abhorrent to me when a fine intelligence is paired with an unsavory character.

To Jakob Laub, May 19, 1909. *CPAE*, Vol. 5, Doc. 161

INTUITION

All great achievements of science must start from intuitive knowledge, namely, in axioms, from which deductions are then made.... Intuition is the necessary condition for the discovery of such axioms.

In Moszkowski, *Conversations with Einstein*, 180

* I believe in intuition and inspiration.... At times I feel certain I am right while not knowing the reason.

"On Science," in *Cosmic Religion* (1931), 97

ITALY AND THE ITALIANS

* The ordinary Italian ... uses words and expressions of a high level of thought and cultural content.... The people of northern Italy are the most civilized people I have ever met.

Quoted by H. Cohen in *Jewish Spectator*, January 1969, 16

* The happy months of my sojourn in Italy are my most beautiful memories.

To Ernesta Marangoni, August 16, 1946. In *Physis* 18 (1976), 174–178

JAPAN AND THE JAPANESE

* The Japanese loves his country and its people more than others do . . . yet he feels like a stranger in foreign countries more than others. I have learned . . . to understand the shyness of the Japanese toward Europeans and Americans: in our countries, education is focused entirely on struggling to survive as individuals. . . . Family bonds are weakened, and . . . isolation of the individual is looked upon as a necessary consequence in the struggle for existence. . . . It is completely different in Japan. The individual here is left to himself much less than in Europe and America. Public opinion here is even greater than in our countries, and sees to it that family structure is not weakened.

> *Kaizo* 5, no. 1 (January 1923), 339

* This flowerlike being: here the ordinary mortal must defer to the poet's words.

> On Japanese women. Ibid.

* May they not forget to keep pure the great heritage that puts them ahead of the West: the artistic configuration of life, the simplicity and modesty of personal needs, and the purity and serenity of the Japanese soul.

> Ibid., 338

* Japan was wonderful. Refined customs, a lively interest in everything, intellectual naiveté but good intelligence—a wonderful people in a picturesque land.

> To Maurice Solovine, Spring 1923. Published in
> *Letters to Solovine*. Einstein was received in Japan
> with great enthusiasm. One reason may be that the
> Japanese characters for "relativity principle" are
> very similar to those for "love" and "sex." See
> Fölsing, *Albert Einstein*, 528

* I have, for the first time, seen a happy and healthy society whose members are fully absorbed in it.

> To Michele Besso, May 24, 1924. Einstein Archive
> 7-349

LIES

He who has never been deceived by a lie does not know the meaning of bliss.

> To Elsa Löwenthal, April 30, 1912. *CPAE*, Vol. 5,
> Doc. 389

LOVE

Love brings much happiness, much more so than pining for someone brings pain.

> To Marie Winteler, his first girlfriend, April 21,
> 1896 (at age 17). *CPAE*, Vol. 1, Doc. 18

* Falling in love is not at all the most stupid thing
that people do—but gravitation cannot be held responsible for it.

> "What Life Means to Einstein," *Saturday Evening
> Post*, October 26, 1929; also quoted in Dukas and
> Hoffmann, *Albert Einstein, the Human Side*, 94

Where there is love, there is no imposition.

> To editor and friend Saxe Commins, Summer 1953.
> Quoted in Sayen, *Einstein in America*, 294

I am sorry that you are having difficulties bringing
your girlfriend [from Dublin to the United States].
But as long as she is there and you are here, you
should be able to maintain a harmonious relationship. So why do you want to press the issue?

> To Cornel Lanczos, February 14, 1955. Einstein
> Archive 15-328

MARRIAGE

My parents . . . think of a wife as a man's luxury
that he can afford only when he is making a comfortable living. I have a low opinion of this view of
the relationship between man and wife, because it
makes the wife and the prostitute distinguishable
only insofar as the former is able to secure a lifelong

contract from the man because of her favorable social rank.

> To Mileva Marić, August 6, 1900. *The Love Letters*, 23; *CPAE*, Vol. 1, Doc. 70

* It is not a lack of real affection that scares me away again and again from marriage. Is it a fear of the comfortable life, of nice furniture, of dishonor that I burden myself with, or even the fear of becoming a contented bourgeois?

> To Elsa Löwenthal, after August 3, 1914. *CPAE*, Vol. 8, Doc. 32

* The solitude and peace of mind are serving me quite well, not the least of which is due to the excellent and truly enjoyable relationship with my cousin; its stability will be guaranteed by the avoidance of marriage.

> To Michele Besso, February 12, 1915. *CPAE*, Vol. 8, Doc. 56

Why should one not admit a man [to the United States] ... who dares to oppose every war except the inevitable one with his own wife?

> In a reply of 1932 to an American women's organization that felt Einstein would be a bad influence on Americans if he visited America. See *Ideas and Opinions*, 7

Marriage is the unsuccessful attempt to make something lasting out of an incident.

> Quoted by Otto Nathan, April 10, 1982, in an
> interview with J. Sayen for *Einstein in America*, 80

* That is dangerous—but then, *any* marriage is dangerous.

> In answer to the question of a Jewish student at
> Princeton about whether interfaith marriage
> should be tolerated. Ibid., 70

Marriage is but slavery made to appear civilized.

> Quoted in Grüning, *Ein Haus für Albert Einstein*, 159

Marriage makes people treat each other as articles of property and no longer as free human beings.

> Ibid.

MATERIALISM

Human beings can attain a worthy and harmonious life only if they are able to rid themselves, within the limits of human nature, of striving to fulfill wishes of the material kind. The goal is to raise the spiritual values of society.

> At a planning conference in Princeton of American
> Friends of Hebrew University. Quoted in the *New
> York Times*, September 20, 1954

MIRACLES

I admit that thoughts influence the body.

Einstein Archive 55-285

* A "miracle" is an exception from lawfulness; hence, where lawfulness does not exist, its exception, i.e., a miracle, also cannot exist.

Quoted and discussed in Jammer, *Einstein and Religion*, 89

MORALITY

One must shy away from questionable undertakings, even when they bear a high-sounding name.

To Maurice Solovine, Spring 1923, on Einstein's resignation from a League of Nations commission. Einstein Archive 21-189; published in *Letters to Solovine*, 59

Morality is of the highest importance—but for us, not for God.

To a banker in Colorado, August 1927. Quoted in Dukas and Hoffmann, *Albert Einstein, the Human Side*, 66

The content of scientific theory itself offers no moral foundation for the personal conduct of life.

In *Forum and Century* 83 (1930), 373

* The destiny of civilized humanity depends more than ever on the moral forces it is capable of generating.

From "Address to the Student Disarmament Meeting," ca. 1930. Published in *Mein Weltbild*; reprinted in *Ideas and Opinions*, 94

There is nothing divine about morality; it is a purely human affair.

Published in *Mein Weltbild* (1934), 18; reprinted in *Ideas and Opinions*, 40

* Without "ethical culture," there is no salvation for humanity.

Published in *Mein Weltbild* (1934), 20

Humanity has every reason to place the proclaimers of high moral standards and values above the discoverers of objective truth. What humanity owes to personalities like Buddha, Moses, and Jesus ranks for me higher than all the achievements of the inquiring constructive mind.

Statement in September 1937. Quoted in Dukas and Hoffmann, *Albert Einstein, the Human Side*, 70

Morality is not a fixed and stark system. . . . It is a task never finished, something that is always present to guide our judgment and inspire our conduct.

From a commencement address at Swarthmore College, Pennsylvania, June 6, 1938. Quoted in the *New York Times*, June 7, 1938

The most important human endeavor is the striving for morality in our actions. Our inner balance and even our very existence depend on it. Only morality in our actions can give beauty and dignity to life.

To a minister in Brooklyn, November 20, 1950. Einstein Archive 28-894, 59–871; quoted in Dukas and Hoffman, *Albert Einstein, the Human Side*, 95

THE MYSTERIOUS

The fairest thing we can experience is the mysterious. It is the fundamental emotion that stands at the cradle of true art and true science. He who does not know it and can no longer wonder, no longer feel amazement, is as good as dead, a snuffed-out candle.

From "What I Believe," *Forum and Century* 84 (1930), 193–194; reprinted in *Ideas and Opinions*, 8–11

* It was a great pleasure for me to tell you about the mysteries with which physics confronts us.

Humans have been given just enough understanding to be able to want to obtain clearer expostula-
tions. The world of mankind would be better if
such humility could be instilled in all.

> To Queen Elizabeth of Belgium, September 19,
> 1932. Einstein Archive 32-353; also quoted in
> Grüning, *Ein Haus für Albert Einstein*, 305

MYSTICISM

* Mysticism is in fact the only criticism people cannot
level against my theory.

> In answer to a Dutch woman, ca. 1921, who had
> met Einstein in the German Embassy in The
> Hague, and who said she liked his mysticism.
> Recounted in Clark, *Einstein*, 340. Though many
> people like to think of Einstein as a mystic, he
> never claimed to have direct subjective communion
> with God or spiritual insight and several times
> expressed his personal aversion to mysticism.

* The mystical trend of our present time, especially
evident in the enthusiastic growth of so-called the-
osophy and spiritualism, is to me a symptom of
confusion and weakness.

> To Lili Halpern-Neuda, February 5, 1921. Einstein
> Archive 43-847

* What I see in Nature is a grand design that we can
understand only imperfectly, one with which a re-

sponsible person must look at with humility. This is a genuine religious feeling and has nothing to do with mysticism.

> Quoted in Dukas and Hoffmann, *Albert Einstein, the Human Side*, 132

PIPE SMOKING

Pipe smoking contributes to a somewhat calm and objective judgment of human affairs.

> Upon accepting life membership in the Montreal Pipe Smokers Club. Quoted in the *New York Times*, March 12, 1950. Einstein was said to be so fond of his pipe that he held on to it even after he fell into the water during a boating accident; see Ehlers, *Liebes Hertz!* 149

POSTERITY

* Dear Posterity: If you have not become more just, more peaceful, and in general more sensible than we are (or were) today, then may the Devil take you! Respectfully expressing his opinion with this devout hope is (or was) your Albert Einstein. Princeton, May 4, 1936.

> Message to posterity written on parchment and placed in an airtight metal box in the cornerstone of the Schuster publishing house (today Simon and Schuster) in New York

PREJUDICE

* Prejudice is part of a tradition which—determined by the events of history—is handed down from generation to generation. One can achieve liberation from prejudice through enlightenment and education. This is a slow and painstaking but purifying process.

> In answer to a questionnaire, asking if U.S. racial prejudice was a symptom of worldwide conflict, October 1948. See Kaller's Autographs catalog, "Jewish Visionaries," 37

THE PRESS

The Press, which is mostly controlled by vested interests, has an excessive influence on public opinion.

> From an interview, *Nieuwe Rotterdamsche Courant*, 1921; also quoted in *Berliner Tageblatt*, July 7, 1921; reprinted in *Ideas and Opinions*, 6

PROHIBITION

Nothing is more destructive of respect for the government and the law of the land than passing laws that cannot be enforced. It is an open secret that the dangerous increase in crime in this country is closely connected with this.

Ibid.

I don't drink, so I couldn't care less.

Statement about Prohibition at a press conference on his arrival in San Diego, December 30, 1930, aboard the *Oakland*. Shown in *Nova*'s Einstein biography, 1979, and in A&E Television's Einstein biography, VPI International, 1991. Einstein disliked alcohol and remained a teetotaler in his later years; see Fölsing, *Albert Einstein*, 81

PSYCHOANALYSIS

I should very much like to remain in the darkness of not having been analyzed.

Response to the suggestion that he undergo Adlerian psychotherapy, January 1927. German psychotherapist H. Freund wanted to study politicians through psychotherapy and had asked Einstein to participate. Quoted in Dukas and Hoffmann, *Albert Einstein, the Human Side*, 35

RICKSHAW PULLERS

* I felt extremely ashamed to be part of such hideous treatment of human beings but couldn't do anything about it. . . . They know how to beseech and beg every tourist until he capitulates.

Travel Diary, October 28, 1922, Colombo, Sri Lanka. Einstein stopped there en route to Singapore, Hong Kong, Shanghai, and Japan.

SAILING

* Sailing in the secluded coves of the coast here is more than relaxing. . . . I have a compass that shines in the dark, like a serious seafarer's. But I am not so talented in this art, and I am satisfied if I can manage to get myself off the sandbanks on which I become lodged.

To Queen Elizabeth of Belgium, March 20, 1954. Einstein Archive 32-410

The sport that demands the least energy.

Quoted by A. P. French, in French, *Einstein: A Centenary Volume*, 61

SCULPTURE

The ability to portray people in still life and in motion requires the highest measure of intuition and talent.

Quoted in Grüning, *Ein Haus für Albert Einstein*, 240

SEX EDUCATION

Regarding sex education: no secrets!

To the World League for Sexual Reform, Berlin,
September 6, 1929. Einstein Archive 48-304; also
quoted in ibid., 306

SUCCESS

Try to become not a man of success, but try rather
to become a man of value.

Quoted in *Life* magazine, May 2, 1955

SUPERSTITION

* By furthering logical thought and a logical attitude,
science can diminish the amount of superstition in
the world.

"On Science," in *Cosmic Religion* (1931), 98

THINKING

Words or language, as they are written or spoken,
do not seem to play any role in my mechanism of
thought.

Quoted in Hadamard, *An Essay on the Psychology of
Invention in the Mathematical Field*, Appendix 2

I vill a little t'ink.

> According to Banesh Hoffmann, this is the phrase
> Einstein used in his broken English when he
> needed more time to think about a problem.
> Quoted in French, *Einstein: A Centenary Volume*,
> 153

I have no doubt that our thinking goes on for the most part without the use of signs (words), and, furthermore, largely unconsciously. For how, otherwise, should it happen that sometimes we "wonder" quite spontaneously about some experience? This "wondering" appears to occur when an experience comes into conflict with a world of concepts that is already sufficiently fixed within us. . . . The development of the world of thinking is in effect a continual flight from wonder.

> From "Autobiographical Notes," in Schilpp, *Albert
> Einstein: Philosopher-Scientist*, 8–9

TOLERANCE

The most important kind of tolerance is tolerance of the individual by society and the state. . . . When the state becomes the dominant element and the individual its weak-willed tool, then all the finer values are lost.

> From an unpublished article on tolerance, 1934.
> Einstein Archive 49-094

TRUTH

The search for truth and knowledge is one of the finest attributes of man—though often it is most loudly voiced by those who strive for it the least.

From "The Goal of Human Existence," broadcast
for the United Jewish Appeal, April 11, 1943.
Einstein Archive 28-587

It is difficult to say what truth is, but sometimes it is so easy to recognize a falsehood.

To Jeremiah McGuire, October 24, 1953. Einstein
Archive 60-483

Whoever is careless with truth in small matters cannot be trusted in important affairs.

From a draft of a television address to be delivered
on occasion of the seventh anniversary of Israel's
independence. Written in April 1955, about a week
before Einstein's death. Quoted in Nathan and
Norden, *Einstein on Peace*, 640

* A pipe-dreaming authority is the worst enemy of truth.

Quoted in Hermann, *Einstein*, 102

VEGETARIANISM

I have always eaten animal flesh with a somewhat guilty conscience.

To Max Kariel, August 3, 1953. Einstein Archive
60-058

When you buy a piece of land to plant your cab-bage and apples, you first have to drain it; that will kill all forms of animal and plant life that exist in that water. Later you would have to kill all the worms and caterpillars etc. that would eat your plants. If you must avoid all this killing on moral grounds, you will in the end have to kill yourself, all for the sake of leaving alive those creatures who have no such conception of higher moral principles.

Ibid. Quoted in *Vegetarisches Universum*, December
1957

* So I am living without fats, without meat, with-out fish, but am feeling quite well this way. It al-most seems to me that man was not born to be a carnivore.

To Hans Muehsam, March 30, 1954. Einstein
Archive 38-435

VIOLENCE

Violence may at times have quickly cleared away an obstruction, but it has never proved itself to be creative.

From "Was Europe a Success?" Quoted in *Einstein on Humanism*, 49

WEALTH

The banal goals of human strivings—possessions, superficial success, luxury—have always seemed contemptible to me.

From "What I Believe," *Forum and Century* 84 (1930), 193–194; reprinted in *Ideas and Opinions*, 8–11

I am absolutely convinced that no amount of wealth can help humanity forward, even in the hands of the most dedicated worker in this cause. The example of great and pure personalities can lead us to noble deeds and views. Money only appeals to selfishness, and, without fail, it tempts its owner to abuse it. Can anyone imagine Moses, Jesus, or Gandhi with the moneybags of Carnegie?

Published in *Mein Weltbild* (1934), 10–11; reprinted in *Ideas and Opinions*, 12–13

The economists will have to revise their theories of value.

> Upon being told that two of his handwritten manuscripts fetched $11.5 million at an auction for the war bonds effort. Recounted by historian Julian Boyd to Dorothy Pratt, February 11, 1944, Princeton University Archives; quoted in Sayen, *Einstein in America*, 150

All I want in my dining room is a pine table, a bench, and a few chairs.

> Quoted in Maja Einstein's biography of her brother; also quoted in Dukas and Hoffmann, *Albert Einstein, the Human Side*, 14

WISDOM

Wisdom is not a product of schooling but of the lifelong attempt to acquire it.

> To an admirer, March 22, 1954. Quoted in Dukas and Hoffmann, *Albert Einstein, the Human Side*, 44

WOMEN

* We men are deplorable, dependent creatures. But compared with these women, every one of us is king, for he stands more or less on his own two feet, not constantly waiting for something outside of

himself to cling to. They, however, always wait for someone to come along who will use them as he sees fit. If this does not happen, they simply fall to pieces.

To Michele Besso, July 21, 1917, in a discussion about Einstein's wife, Mileva. *CPAE*, Vol. 8, Doc. 239

Very few women are creative. I would not send a daughter of mine to study physics. I'm glad my wife doesn't know any science. My first wife did.

Quoted by Esther Salaman, who had been a young student in Berlin when Einstein was there, in the *Listener*, September 8, 1968; also quoted in Highfield and Carter, *The Private Lives*, 158

As in all other fields, in science the way should be made easy for women. Yet it must not be taken amiss if I regard the possible results with a certain amount of skepticism. I am referring to certain restrictive parts of a woman's constitution that were given her by Nature and which forbid us from applying the same standard of expectation to women as to men.

In Moszkowski, *Conversations with Einstein*, 79

When women are in their homes, they are attached to their furnishings . . . they are always fussing with them. When I am with a woman on a trip, I am the only piece of furniture she has available, and she cannot refrain from circling around me all day and making some improvements on me.

Quoted in Frank, *Einstein: His Life and Times*, 126

WORK

Work is the only thing that gives substance to life.

To son Hans Albert, January 4, 1937. Einstein Archive 75-926

It is really a puzzle what drives one to take one's work so devilishly seriously. For whom? For oneself? One soon departs this world, after all. For one's companions? For posterity? *No*. It remains a puzzle.

To artist Joseph Scharl, December 27, 1949. Einstein Archive 34-207

I am also convinced that one gains the purest joy from spiritual things only when they are not tied in with earning one's livelihood.

To L. Manners, March 19, 1954. Einstein Archive 60-401

YOUTH

* Truly novel ideas emerge only in one's youth. Later on one becomes more experienced, famous—and foolish.

> To Heinrich Zangger, December 6, 1917. *CPAE*,
> Vol. 8, Doc. 403

* When I read your letters, I am very much reminded of my youth. In one's thoughts, one tends to set oneself against the world. One compares one's strengths with everything else, one alternates between despondency and self-assurance. One has the feeling that life is eternal and that everything one does and thinks is so important.

> To son Eduard, December 27, 1932. Quoted in
> Rosenkranz, *Albert through the Looking-Glass*, 17

O, Youth: Do you know that yours is not the first generation to yearn for a life full of beauty and freedom? Do you know that all your ancestors have felt the same as you do—and fell victim to trouble and hatred? Do you know also that your fervent wishes can only find fulfillment if you succeed in attaining a love and an understanding of people, and animals, and plants, and stars, so that every joy becomes your joy and every pain your pain?

> Written into a neighbor's autograph album in Caputh,
> Germany, 1932. Quoted in Dukas and Hoffmann,
> *Albert Einstein, the Human Side*, 129

Einstein with an Einstein puppet. (Photo by Harry Burnett/Courtesy Caltech Archives)

Over the past four years, a number of people have sent me the sources of quotations that appeared in this section in the last edition. I thank them for this help, and I have inserted the quotations into the text in the appropriate sections. I am still looking for the sources of many of the following quotations. Some sound genuine, some are apocryphal, and others are no doubt fakes, created by those who wanted to use Einstein's name to lend credibility to a cause or an idea. Hundreds can be found on the Internet, on calendars, and in little books containing undocumented quotations, but I include here only quotations sent to me by curious readers.

MISATTRIBUTED TO EINSTEIN

International law exists only in textbooks on international law.

Actually said by Ashley Montagu in his interview with Einstein. See Montagu's "Conversations with Einstein," *Science Digest*, July 1985

* Education is that which remains, if one has forgotten everything he learned in school.

Not originally said by Einstein, though he agreed with it. He quoted this passage by an anonymous "wit" in "On Education," in *Out of My Later Years*, 38

* We use only 10% of our brains.

> This myth is repeated frequently and is false.
> Several articles have been written about it and
> were sent to me by readers.

* As a young man, my fondest dream was to become
a geographer. However, while working in the customs office, I thought deeply about the matter and
concluded that it was far too difficult a subject.
With some reluctance, I turned to physics as a
substitute.

> Sent by a member of a geography department
> in South Dakota, who said this quotation is
> circulating around many geography departments.
> Einstein was obviously already a physicist by the
> time he worked in the Patent (not Customs) Office,
> so it was a bit late to be thinking of geography as
> a possible future career.

POSSIBLY BY EINSTEIN

* Everything should be made as simple as possible,
but not simpler.

> This quotation prompts the most queries.
> Everyone seems to know it, yet no one can find its
> source. There appear to be several variants of it,
> too, most commonly, "A *theory* should be made
> as simple as possible, but not simpler." Even
> journalist William Safire of the *New York Times*
> *Magazine* quoted it (without documentation, for

which I reproached him) in a piece he wrote about Occam's Razor. Occam's Razor, also known as the "principle of parsimony," is a scientific and philosophic rule that says that entities should not be multiplied unnecessarily. This is interpreted as requiring that the simpler of competing theories be preferred to the more complex, or that explanations of unknown phenomena be sought first in terms of known quantities. It could be that someone long ago misattributed Occam's Razor to Einstein, and that this saying is actually over six hundred years older than we think it is. (William of Ockham lived ca. 1285–1349.) It is also known that Einstein was sometimes not completely original and borrowed from other thinkers, substituting his own variants to suit the occasion. Or, the quotation may be a paraphrase of some of Einstein's other statements about simplicity.

* It is not enough for a handful of experts to attempt the solution of a problem, to solve it, and then apply it. The restriction of knowledge to an elite group destroys the spirit of society and leads to its intellectual impoverishment.

> This may be from an address at Caltech, 1931.
> Needs to be verified.

* The substance of our knowledge resides in the detailed terminology of a field.

No amount of experimentation can ever prove me right; a single experiment can prove me wrong.

Everything that is really great and inspiring is created by the individual who can labor in freedom.

In science the work of the individual is so bound up with that of his scientific predecessors and contemporaries that it appears almost as an impersonal product of his generation.

The effort to get at the truth has to precede all other efforts.

In any conflict between humanity and technology, humanity will win.

Few people are capable of expressing with equanimity opinions that differ from the prejudices of their social environment.

* A human being is a part of the whole called by us "the universe," a part limited in time and space. He experiences himself, his thoughts and feelings, as something separate from the rest—a kind of optical illusion of his consciousness. This delusion is a kind of prison for us, restricting us to our personal desires and affection for a few persons nearest to us. Our task must be to free us from this prison by widening our circle of understanding and compassion to embrace all living creatures and the whole of nature in its beauty.

* Science as something already in existence, already completed, is the most objective, impersonal thing that we humans know. Science as something coming into being, as a goal, is just as subjectively, psychologically conditioned as are all other human endeavors.

* It is in fact nothing short of a miracle that modern methods of instruction have not yet entirely strangled the holy curiosity of inquiry; for this delicate little plant, aside from stimulation, stands mainly in need of freedom; without this it goes to wrack and ruin without fail.

* It is possible that there exist emanations that are still unknown to us. Do you remember how electrical currents and "unseen waves" were laughed at? The knowledge about man is still in its infancy.

* Scientific truth is nothing but conditional truth.

* The significant problems we face cannot be solved at the same level of thinking we were at when we created them. (Variant: The world we have created today as a result of our thinking thus far has problems which cannot be solved by thinking the way we thought when we created them.)

  Another common query. Perhaps it's a paraphrase.

* Before long, mankind must establish without fail a world leader for the sake of true peace. The person who becomes the world leader cannot be concerned with either military or monetary power. He must come from the oldest country that transcends the history of all countries and that has a noble national character. World culture began in Asia and must return to Asia, that is, to Asia's highest peak, Japan. We are grateful to God for this. Heaven created such a noble country, Japan, for us.

> Allegedly said by Einstein in a speech at
> Tohoku University in Sendai, 1922. Needs to
> be verified.

PROBABLY NOT BY EINSTEIN

Nothing will benefit human health and increase the chances for survival on Earth as much as the evolution of a vegetarian diet.

* Truth is what withstands the test of experience.

* Not everything that counts can be counted, and not everything that can be counted counts.

* If you think intelligence is dangerous, try ignorance.

* There is no hope for an idea that at first does not seem insane.

* The religion of the future will be a cosmic religion.

* We should take care not to make intellect our god; it has, of course, powerful muscles, but no personality.

* Common sense is the collection of prejudices acquired by age eighteen.

The discovery of a nuclear chain reaction need not bring about the destruction of mankind any more than the discovery of matches.

* Two things inspire me to awe—the starry heavens above and the moral universe within.

* The probability of life originating from accident is comparable to the probability of the unabridged dictionary resulting from an explosion in a print shop. (Variant: The idea that this universe in all its millionfold order and precision is the result of blind chance is as credible as the idea that if a print shop blew up all the type would fall down again in the finished and faultless form of the dictionary.)

* If the facts don't fit the theory, change the facts.

* There are only two ways to live your life. One is as though nothing is a miracle. The other is as though everything is a miracle.

* Compounded interest is more complicated than relativity theory. (Variant: The most powerful force in the universe is compound interest.)

> Quoted by various economists, including Burton
> Malkiel in the *Princeton Spectator*, May 1997, and
> on several financial Web sites on the Internet.
> (Thanks to Steven Feldman for showing me the
> variant.)

* The most significant invention of the nineteenth century is compounded interest.

* The most difficult thing to understand is the income tax. (Variant: The hardest thing to understand in the world is the income tax.)

> Quoted in the *Macmillan Book of Business and
> Economics Quotes* by M. Jackson (1984). No source
> given.

* Preparing a tax return is more complicated than relativity theory.

* If the bee becomes extinct, mankind will have only four years to live: no bees, no pollination, no plants, no animals, no humans.

* As the circle of light increases, so does the circumference of darkness.

* Astrology is a science in itself and contains an illuminating body of knowledge. It taught me many

things and I am greatly indebted to it. Geophysical evidence reveals the power of the stars and the planets in relation to the terrestrial. This is why astrology is like a life-giving elixir to mankind.

> An excellent example of a quotation someone
> made up and attributed to Einstein in order to lend
> an idea credibility. Yet several people have asked
> me to confirm it.

* The practice of science as a whole finds truths and leads to a correct understanding of the universe, but science in actual practice is riddled with mistakes and the residue of human frailties.

* So long as you pray to God to ask him for *something*, you are not a religious person.

* In the middle of every difficulty lies opportunity.

> A truism that is probably far older than Einstein.
> See also next page.

* The lowest level of awareness is "I know." Then, "I don't know," "I know I don't know," "I don't know that I don't know."

* The levels of intelligence are "Smart, intelligent, brilliant, genius, *simple*."

* Death means that one can no longer listen to Mozart.

* Einstein's "Rules of Work"

1. Out of clutter, find simplicity.
2. From discord, find harmony.
3. In the middle of difficulty lies opportunity.

The first "rule" is probably a paraphrase of
Einstein's many quotations about the value of
simplicity. I've traced the second rule to Horace,
the Roman poet and satirist, who had it as
"Concordia discors" (harmony in discord) in his
Epistles I, xii.19. And the third rule has probably
been in general use for ages.

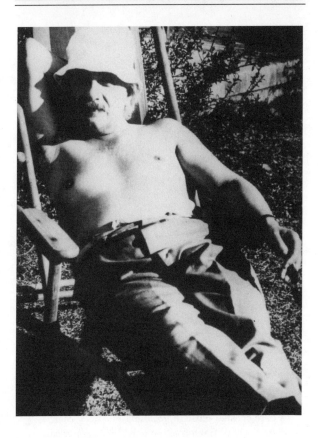

A bare-chested Einstein, reclining in a deck chair, Palm Springs, California, January 1932. (American Stock/Archive Photos)

Tell me what to do if he says yes. I had to offer the post to him because it is impossible not to. But if he accepts, we are in for trouble.

David Ben Gurion to Yitzak Navon, after Israeli ambassador to the United States, Abba Eban, was instructed to offer the presidency of Israel to Einstein in November 1952. Quoted in Holton and Elkana, *Albert Einstein: Historical and Cultural Perspectives*, 295

When something struck him as funny, his eyes twinkled merrily and he laughed with his whole being. . . . He was ready for humor.

Algernon Black, 1940. Einstein Archive 54-834

Through Albert Einstein's work the horizon of mankind has been immeasurably widened, at the same time as our world picture has attained a unity and harmony never dreamed of before. The background for such an achievement was created by preceding generations of the world community of scientists, and its full consequences will only be revealed to coming generations.

Niels Bohr. Quoted in Einstein's obituary in the *New York Times*, April 19, 1955

Einstein would be one of the greatest theoretical physicists of all time even if he had not written a single line on relativity.

> Max Born. Quoted in Hoffmann, *Albert Einstein: Creator and Rebel*, 7

He always took his celebrity with humor and laughed at himself.

> Family friend Thomas Bucky, shown in A&E Television's Einstein biography, VPI International, 1991

* They cheer me because they all understand me, and they cheer you because no one understands you.

> Charlie Chaplin, after the premiere of *City Lights* in Los Angeles in January 1931, to which Chaplin had invited Einstein. See Fölsing, *Albert Einstein*, 457

* His short skull seems unusually broad. His complexion is matte light brown. Above his large sensuous mouth is a thin black mustache. The nose is slightly aquiline. His striking brown eyes radiate deeply and softly. His voice is attractive, like the vibrant note of a cello.

> Einstein's student Louis Chavan. Quoted in Max Flückiger, *Albert Einstein in Bern* (1972), 11–12

The contrast between his soft speech and ringing laughter was enormous. . . . Every time he made a point he liked, or heard something that appealed to him, he would burst into a booming laughter that would echo from wall to wall.... I had been prepared to know what he would look like ... but I was totally unprepared for this roaring, booming, friendly, all-enveloping laughter.

> Bernard Cohen. Quoted in an interview with
> Whitrow, in Whitrow, *Einstein*, 83

I was able to appreciate the clarity of his mind, the breadth of his information, and the profundity of his knowledge.... One has every right to build the greatest hopes on him and to see in him one of the leading theoreticians of the future.

> Marie Curie, 1911. Quoted in Hoffmann, *Albert
> Einstein: Creator and Rebel*, 98–99

* Your mere presence here undermines the class's respect for me.

> Said to Einstein by his seventh-grade teacher,
> Dr. Joseph Degenhart, who also predicted that he
> "would never get anywhere in life." In a draft
> letter to Philipp Frank, 1940; see also *CPAE*, Vol. 1,
> lxiii

The professor never wears socks. Even when he was invited by Mr. Roosevelt to the White House he didn't wear socks.

> Helen Dukas. Related by Philippe Halsman in
> French, *Einstein: A Centenary Volume*, 27

Einstein was prone to talk about God so often that
I was led to suspect he was a disguised theologian.

> Writer Friedrich Dürrenmatt. In *Albert Einstein: Ein
> Vortrag*, 12

* After a careful study of the plates I am prepared to
say that there can be no doubt that they confirm
Einstein's prediction. A very definite result has
been obtained that light is deflected in accordance
with Einstein's law of gravitation.

> The British Astronomer-Royal, Sir Frank Dyson,
> after the Dyson-Eddington expedition in the midst
> of World War I had confirmed Einstein's general
> theory of relativity. *Observatory* 32 (1919), 391

* Here we have nothing but people who love *you* and
not just your cerebral cortex.

> Close friend Paul Ehrenfest. In a letter to Einstein,
> September 8, 1919

* God has put so much into him that is beautiful, and
I find him wonderful, even though life at his side is
debilitating and difficult in every respect.

> Elsa Einstein. In a letter to Hermann Struck and
> wife, 1929. Quoted in Fölsing, *Albert Einstein*, 429

It is not ideal to be the wife of a genius. Your life
does not belong to you. It seems to belong to every-

one else. Nearly every minute of the day I give to my husband, and that means to the public.

Elsa Einstein. Quoted in her obituary in the *New York Times*, December 22, 1936, two days after her death

* Oh, my husband does that on the back of an old envelope!

Elsa Einstein, after a host at Mount Wilson Observatory in California explained to her that the giant telescope is used to find out the shape of the universe. Reported by Bennett Cerf in *Try and Stop Me* (New York: Simon and Schuster, 1944)

Probably the only project he ever gave up on was me. He tried to give me advice, but he soon discovered that I was too stubborn and that he was just wasting his time.

Hans Albert Einstein, *New York Times*, July 27, 1973. Quoted in Pais, *Einstein Lived Here*, 199

He was very fond of nature. He did not care for large, impressive mountains, but he liked surroundings that were gentle and colorful and gave one lightness of spirit.

Hans Albert Einstein. Quoted in an interview with Bernard Mayer, in Whitrow, *Einstein*, 21

* Every once in a while . . . he spanked me, just like anyone else would do.

Ibid.

He often told me that one of the most important things in his life was music. Whenever he felt he had come to the end of the road or into a difficult situation in his work, he would take refuge in music and that would usually resolve all his difficulties.

Ibid.

* His work habits were rather strange. . . . Even when there was a lot of noise, he could lie down on the sofa, pick up a pen and paper, precariously balance an inkwell on the backrest, and engross himself in a problem so much so that the background noise stimulated rather than disturbed him.

Maja Einstein. *CPAE*, Vol. 1, lxiv

When one was with him on the sailboat, you felt him as an element. He had something so natural and strong in him because he was himself a piece of nature. . . . He sailed like Odysseus.

Margot Einstein, May 4, 1978. In an interview with J. Sayen, in Sayen, *Einstein in America*, 132

No other man contributed so much to the vast ex-
pansion of twentieth-century knowledge.

Statement by President Dwight D. Eisenhower
upon Einstein's death. Quoted in Einstein's
obituary in the *New York Times*, April 19, 1955

* In view of his radical background, this office would
not recommend the employment of Dr. Einstein, on
matters of a secret nature, without a very careful
investigation, as it seems unlikely that a man of his
background could, in such a short time, become a
loyal American citizen.

Recommendation by the Federal Bureau of
Investigation (FBI), which had never been
informed of Einstein's letter to President Roosevelt
warning him about the possibility of the Germans'
building a bomb. Quoted by Richard Schwartz in
Isis 80 [1989], 281–284

Einstein's conversation was often a combination of
inoffensive jokes and penetrating ridicule so that
some people could not decide whether to laugh or
to feel hurt. . . . Such an attitude often appeared to
be an incisive criticism, and sometimes even created an impression of cynicism.

Philipp Frank. In Frank, *Einstein: His Life and
Times*, 77

He, who had always had something of a bohemian in him, began to lead a middle-class life ... in a household such as was typical of a well-to-do Berlin family. ... When one entered ... one found that Einstein still remained a "foreigner" in such a surrounding—a bohemian guest in a middle-class home.

> Ibid., 124

* He is cheerful, assured, and courteous, understands as much of psychology as I do of physics, and so we had a pleasant chat.

> Sigmund Freud, 1926, on a visit to Berlin, where he
> met Einstein. In a letter to S. Ferenczi, January 2,
> 1927, in *The Collected Papers of Sigmund Freud*,
> ed. Ernest Jones (out of print)

* [I have finished writing] the tedious and sterile so-called discussion with Einstein.

> Sigmund Freud to Max Eitinger, September 8, 1932,
> on Freud and Einstein's correspondence published
> by the League of Nations in 1933 as *Why War?*
> In ibid., 175

* Einstein's fiddling was like that of a lumberjack.

> Comment by professional violinst Walter
> Friedrich. Quoted in Herneck, *Einstein Privat*
> (Berlin, 1978), 129

Of course, the old man agrees with almost anything nowadays.

Cosmologist George Gamow. Written on the
bottom of a letter from Einstein of August 4, 1948,
in which Einstein states that one of Gamow's ideas
is probably correct. In Reines, *Cosmology, Fusion
and Other Matters*, 310

A man distinguished by his desire, if possible, to efface himself and yet impelled by the unmistakable power of genius which would not allow the individual of whom it had taken possession to rest for one moment.

Lord Haldane. Quoted in *The Times* (London),
June 14, 1921

The essence of Einstein's profundity lay in his simplicity; and the essence of his science lay in his artistry—his phenomenal sense of beauty.

Banesh Hoffmann. In Hoffmann, *Albert Einstein:
Creator and Rebel*, 3

Einstein, with his feelings of humility, awe, and wonder, and his sense of oneness with the universe, belongs with the great religious mystics.

Ibid., 94. Einstein, however, did not consider
himself a mystic.

When it became clear that [we could not solve a problem], Einstein would stand up quietly and say, in his quaint English, "I vill a little t'ink." So saying he would pace up and down or walk around in circles, all the time twirling a lock of his long, graying hair around his finger.

Banesh Hoffmann. Recollection quoted in
Whitrow, *Einstein*, 75

The "Great Relative."

Name given Einstein by Hopi Indians on his visit
to the United States, 1921. Recounted in A&E
Television's Einstein biography, VPI International,
1991

Einstein gave his wife the greatest care and sympathy. But in this atmosphere of approaching death, Einstein remained serene and worked constantly.

Leopold Infeld, on Einstein's coping with wife
Elsa's terminal illness of heart and kidney disease.
In Infeld, *The Quest*, 282

The greatness of Einstein lies in his tremendous imagination, in the unbelievable obstinacy with which he pursues his problems.

Ibid., 208

If Einstein were to enter your room at a party and he were introduced to you as a "Mr. Eisenstein" of whom you knew nothing, you would still be fascinated by the brilliance of his eyes, by his shyness and gentleness, by his delightful sense of humor, by the fact that he can twist platitudes into wisdom. . . . You feel that before you is a man who thinks for himself. . . . He believes what you tell him because he is kind, because he wishes to be kind, and because it is much easier to believe than to disbelieve.

> Leopold Infeld. In Infeld, *Albert Einstein*, 128

With all his phenomenal intellect, he is still a naïve and altogether spontaneous human being.

> Erich Kahler, 1954. Einstein Archive 38-279

* You'd better watch out, you'd better take care, Albert says *E* equals *m c* square.

> From the song "Einstein A-go-go," by the pop group Landscape

* It is as important an event as would be the transfer of the Vatican from Rome to the New World. The pope of physics has moved, and the United States will now become the center of the natural sciences.

> Paul Langevin, on Einstein's move to America. Quoted in Pais, *A Tale of Two Continents*, 227

* Anyone who has had the pleasure of being close to Einstein knows that he is not surpassed by anyone in respecting the intellectual property of others, in personal modesty, and in his distaste for publicity.

> Max von Laue, Walther Nernst, and Heinrich Rubens, in a joint declaration of support for Einstein as anti-Semitism and anti-relativism were spreading among German physicists. *Tägliche Rundschau*, August 26, 1920

Jewish physics can best and most justly be characterized by recalling the activity of one who is probably its most prominent representative, the pureblooded Jew, Albert Einstein. His relativity theory was supposed to transform all of physics, but when faced with reality it did not have a leg to stand on. In contrast to the intractable and solicitous desire for truth in the Aryan scientist, the Jew lacks to a striking degree any comprehension of truth.

> German physicist and winner of the 1905 Nobel Prize, Philip Lenard, in his book, *German Physics* (Munich: Lehmann's Verlag, 1936). Earlier in the century, Lenard and Einstein had had great respect for each other, then came into conflict over the general theory of relativity. Lenard's experiments on the photoelectric effect had led Einstein toward the hypothesis of light quanta.

* It was interesting to see them together—Tagore, the poet with the head of a thinker, and Einstein, the thinker with the head of a poet. It seemed to an ob-

server as though two planets were engaged in a chat.

> Journalist Dmitri Marianoff, Margot Einstein's husband, to the *New York Times*, on his observations of a conversation between Einstein and Indian poet, musician, and mystic Rabindranath Tagore on July 14, 1930. See Tagore, "Farewell to the West" (1930–1931), 294–295; idem, *The Religion of Man* (New York: Macmillan, 1931), Appendix 2, 221–225

Anyone who advises Americans to keep secret information which they may have about spies and saboteurs is himself an enemy of America.

> Senator Joseph McCarthy, regarding Einstein's advocacy of refusing to testify at the House Un-American Activities Committee hearings. *New York Times*, June 14, 1953

* He wore his usual jersey, baggy pants, and slippers. What especially struck me as he approached the doorway was that he seemed not to walk but to glide in a sort of undeliberate dance. It was enchanting. And there he was, bright, sad eyes, cascading white hair, with a smile of greeting on his face, a firm handshake.

> From anthropologist Ashley Montagu's "Conversations with Einstein," *Science Digest*, July 1985

* When he offered his last important work to the publishers, he warned them that there were no more than twelve persons in the whole world who would understand it, but the publishers took the risk.

> This false report by a *New York Times* reporter, November 10, 1919, regarding the general theory of relativity, became established in the Einstein mythology. On December 3, 1919, another *New York Times* reporter asked if this statement was true, upon which "the doctor laughed good-humoredly." See Fölsing, *Albert Einstein*, 447, 451

"For his unique services to theoretical physics and in particular for his discovery of the photoelectric effect."

> Nobel Prize Committee, official citation for the Nobel Prize in Physics, 1921. Note that no mention was made of relativity theory, which was still a controversial topic at the time. Einstein had been nominated for the prize each year from 1910 to 1918 except for 1911 and 1915. See Pais, *Subtle Is the Lord*, 505; see also my discussion of "The Nobel Prize" in the next section

He has a quiet way of walking, as if he is afraid of alarming the truth and frightening it away.

> Japanese cartoonist Ippei Okamoto on Einstein's visit to Japan, November 1922. See manuscript, "Einstein's 1922 Visit to Japan," in Einstein Archive 36-409

He was almost wholly without sophistication and wholly without worldliness. . . . There was always in him a powerful purity at once childlike and profoundly stubborn.

Robert Oppenheimer. In "On Albert Einstein,"
New York Review of Books, March 17, 1966

He responded with one of the most extraordinary kinds of laughter. . . . It was rather like the barking of a seal. It was a happy laughter. From that time on, I would save a good story for our next meeting, for the sheer pleasure of hearing Einstein's laugh.

Abraham Pais. Quoted in Bernstein, *Einstein*, 77

What Einstein said wasn't all that stupid.

Wolfgang Pauli as a student, after hearing
Einstein, twenty years his senior, give a lecture.
Quoted in Ehlers, *Liebes Hertz!* 47

Doctor with the bushy head
Tell us that you're not a Red.
Tell us that you do not eat
Capitalists in the street.
Say to us it isn't true
You devour their children too.
Speak, oh speak, and say you're notsky
Just a bent-space type of Trotsky.

Verse written by popular newspaper columnist
H. I. Phillips during the McCarthy era, poking fun

at anti-Communist opposition to Einstein's admittance to the United States two decades earlier. Quoted by Norman F. Stanley in *Physics Today*, November 1995, 118

* In boldness it exceeds anything so far achieved in speculative natural science, in philosophical cognition theory. Non-Euclidean geometry is child's play by comparison.

Max Planck on Einstein's definition of time, in a lecture delivered at Columbia University, Spring 1909 (published Leipzig, 1910, 117ff)

* Even though in political matters a deep gulf divides us, I am also absolutely certain that in the centuries to come Einstein will be celebrated as one of the brightest stars that ever shined on our academy.

Max Planck to Heinrich von Ficker, March 31, 1933, on Einstein's resignation from the Prussian Academy of Sciences. Quoted in Christa Kirstin and H.-J. Treder, *Albert Einstein in Berlin, 1913– 1933* (Berlin, 1979)

Einstein loved women, and the commoner and sweatier and smellier they were, the better he liked them.

Peter Plesch, quoting his father, János. In Highfield and Carter, *The Private Lives of Albert Einstein*, 206

What we must particularly admire in him is the facility with which he adapts himself to new concepts and that he knows how to draw from them every conclusion.

> Henri Poincaré, 1911. Quoted in Hoffmann, *Albert Einstein: Creator and Rebel*, 99

* Einstein, a genius in science, is weak, indecisive, and contradictory outside his own field.... His continuous change of opinion and . . . change in his actions are worse than the inflexible obstinacy of a declared enemy.

> Pacifist Romain Rolland, diary entry of September 1933. Quoted in Nathan and Norden, *Einstein on Peace*, 233

* To Einstein, hair and violin,
  We give our final nod.
  He's understood by just two folks:
  Himself . . . and sometimes God.

> An ode by Jack Rossetter. Sent by a reader from India

Einstein was indisputably one of the greatest men of our time. He had, in a high degree, the simplicity characteristic of the best men in science—a simplicity which comes of a single-minded desire to know and understand things that are completely impersonal.

Bertrand Russell. In *The New Leader*, May 30, 1955

He removed the mystery from gravitation, which everybody since Newton had accepted, with a reluctant feeling, as unintelligible.

Bertrand Russell. Quoted in Whitrow, *Einstein*, 22

Of all the public figures that I have known, Einstein was the one who commanded my most wholehearted admiration.... Einstein was not only a great scientist, he was a great man. He stood for peace in a world drifting towards war. He remained sane in a mad world, and liberal in a world of fanatics.

Ibid., 90

* For heaven's sake, Albert, can't you *count*?

Pianist Artur Schnabel, after Einstein made several wrong entrances in a quartet rehearsal. Schnabel was teaching a master class at Princeton at the time. Recalled by Mike Lipskin and quoted by Herb Caen in the *San Francisco Chronicle*, February 3, 1996

Even though without writing each other, we are in mental communication; for we respond to our dreadful times in the same way and tremble together for the future of mankind. . . . I like it that we have the same given name.

Albert Schweitzer. In a letter to Einstein, February 20, 1955. Einstein Archive 33-236

Tell Einstein that I said the most convincing proof I can adduce of my admiration for him is that his is the only one of these portraits [of celebrities] I paid for.

George Bernard Shaw. Recalled by Archibald Henderson in the *Durham Morning Herald*, August 21, 1955. Einstein Archive 33-257. Einstein's reply: "That is very characteristic of Bernard Shaw, who has declared that money is the most important thing in the world."

Ptolemy made a universe, which lasted 1400 years. Newton, also, made a universe, which lasted 300 years. Einstein has made a universe, and I can't tell you how long that will last.

George Bernard Shaw, at a banquet in England honoring Einstein. Quoted in Cassidy, *Einstein and Our World*, 1. Also shown in *Nova*'s Einstein biography, 1979. Another version exists in Shaw's "An Appreciation," in Einstein's *Cosmic Religion*, 32–33, 38: "Only three [men] made universes. Newton made a universe which lasted 300 years. Einstein has made a universe, which I suppose you want me to say will never stop, but I don't know how long it will last. . . . I rejoice in the new universe Einstein has produced."

* There is only one fault with his cosmic religion: he put an extra letter in the word—the letter "s."

    Bishop Fulton J. Sheen. Quoted in Clark, *Einstein*, 517

* What did surprise me was his physique. He had come in from sailing and was wearing nothing but a pair of shorts. It was a massive body, very heavily muscled; he was running to fat round the midriff and in the upper arms, rather like a footballer in middle age, but he was an unusually strong man.

    C. P. Snow on his visit to Einstein in 1937. Quoted by Richard Rhodes, *The Making of the Atom Bomb* (New York: Touchstone Books, 1995)

To me, he appears as out of comparison the greatest intellect of this century, and almost certainly the greatest personification of moral experience. He was in many ways different from the rest of the species.

    C. P. Snow. In "Conversations with Einstein," as quoted in French, *Einstein: A Centenary Volume*, 193

He was a Zionist on general humanitarian grounds rather than on nationalistic grounds. He felt that Zionism was the only way in which the Jewish problem in Europe could be settled. . . . He was never in favor of aggressive nationalism, but he felt that a Jewish homeland in Palestine was essential to save the remaining Jews in Europe. . . . After the State of Israel was established, he said that somehow he felt happy he was not there to be involved in the deviations from the high moral tone he detected.

Ernst Straus. Quoted in Whitrow, *Einstein*, 87–88

* Nobody in *football* should be called a genius. A genius is a guy like Norman Einstein.

Football commentator and former player Joe Theisman. Quoted in *The Book of Truly Stupid Sports Quotes* (New York: HarperCollins, 1996)

One of the greatest—perhaps *the* greatest—of achievements in the history of human thought.

Joseph John Thomson, discoverer of the electron, referring to Einstein's work on general relativity, 1919. Quoted in Hoffmann, *Albert Einstein: Creator and Rebel*, 132

* He had the kind of male beauty that, especially at the beginning of the century, caused great commotion.

Quoted by Vallentin, *Das Drama Albert Einsteins*, 9

* [Einstein] acted on women as a magnet acts on iron filings.

Konrad Wachsmann, the architect of Einstein's house in Caputh. Quoted in Grüning, *Ein Haus für Albert Einstein*, 158

During our crossing, Einstein explained his theory
to me every day, and by the time we arrived I was
fully convinced he understood it.

> Chaim Weizmann, Spring 1921, after he escorted
> Einstein to the United States on the SS *Rotterdam*
> on behalf of a Zionist delegation. Quoted in Seelig,
> *Helle Zeit, dunkle Zeit*, 136

* [Einstein] is acquiring the psychology of a prima
donna who is beginning to lose her voice.

> Chaim Weizmann, 1933, in response to Einstein's
> requests for reforms at Hebrew University.
> Quoted in Norman Rose, *Chaim Weizmann* (New
> York, 1986), 297

Einstein was a physicist and not a philosopher.
But the naïve directness of his questions was
philosophical.

> C. F. von Weizsaecker. Quoted in Aichenburg and
> Sexl, *Albert Einstein*, 159

We salute the new Columbus of science voyaging
through the strange seas of thought.

> Dean Andrew Fleming West of Princeton
> University, after reading a citation before
> President John Grier Hibben conferred an
> honorary doctorate on Einstein, May 9, 1921.
> Quoted in Alexander Leitch, *A Princeton
> Companion* (Princeton University Press, 1978),
> 153. (I had originally attributed this quotation to

Hibben, not Fleming, based on information in
Philipp Frank's biography.)

* He is not a good teacher for mentally lazy gentle-
men who merely want to fill up a notebook and
then learn it by heart for an exam; he is not a
smooth talker. But anyone who wants to learn how
to construct physical ideas, carefully examine all
premises, take note of the pitfalls and problems, review the reliability of his reflections, will find Einstein a first-rate teacher.

Heinrich Zangger, in a letter to Ludwig Ferrer, October 9, 1911, recommending Einstein for a post at
the ETH in Zurich

Einstein's [violin] playing is excellent, but he does
not deserve his world fame; there are many others
just as good.

A Berlin music critic on an early 1920s perfor-
mance, unaware that Einstein's fame derived from
physics, not music. Quoted in Reiser, *Albert Einstein*, 202–203

"Prof. Einstein's Got a New Baby: Formula Keeps
Our Man Up Nights."

Headline of a book review of *The Meaning of Relativity* which appeared in the *Daily Mirror* (New
York), March 30, 1953. It referred to the appendix
published two years before Einstein's death in

which the physicist presented a greatly simplified
derivation of the equations of general relativity.
(Contributed by Trevor Lipscombe)

* Here lies Einstein, an enterprising Teuton
  Who, relatively speaking, silenced Newton.

> Epitaph for Einstein by an unidentified author.
> Quoted by Ashley Montagu in "Conversations
> with Einstein," *Science Digest*, July 1985. Einstein
> would probably have taken issue with the
> "Teuton" characterization.

* Three wonderful people called Stein;
  There's Gert and there's Ep and there's Ein.
  Gert writes in blank verse,
  Ep's sculptures are worse,
  And nobody understands Ein.

> Verse by an unidentified author. Quoted in ibid.

And, finally:

All boys are idiots except for Albert Einstein.

> Mary Lipscombe, age 8, to Lottie Appel, age 6,
> daughters of my colleagues Trevor and Fred,
> respectively. Overheard at Princeton University
> Press Christmas party, December 21, 1999

Answers to the
Most Common
Nonscientific Questions
about Einstein

Einstein at home with stepdaughter Margot (*center*), Helen Dukas, and pet dog Chico, March 1953. (Esther Bubley/Archive Photos)

The following information has been culled from various sources in the Einstein Archive and in the published literature. Much of it can be found in the standard biographies of Einstein, such as those by Albrecht Fölsing, Abraham Pais, and Jamie Sayen, an Einstein neighbor.

PHYSICISTS EINSTEIN ADMIRED THE MOST

Michael Faraday, H. A. Lorentz, James Clerk Maxwell, and Isaac Newton.

PHILOSOPHERS WHO INFLUENCED HIM THE MOST

David Hume, for his criticism of traditional assumptions and dogmas; Ernst Mach, for his criticism of Newton's ideas concerning space, for his critical examination of Newtonian mechanics, and for his encouragement of intellectual skepticism; Baruch (Benedict de) Spinoza, for his views on religion; and Arthur Schopenhauer, for the inspirational "A man can do what he wants, but not want what he wants." (See Frank, *Einstein: His Life and Times*, 52; Whitrow, *Einstein*, 12-13; *Ideas and Opinions*, 8)

BOOKS AND AUTHORS HE ENJOYED

Gandhi's autobiography; John Hersey's *A Bell for Adano* and *The Wall*; Thornton Wilder's *Our Town*; books by Dostoevsky, Tolstoy, and Herodotus; Spinoza's writings on religion. Books on science that he recommended in 1920 were Hermann Weyl's *Time, Space, and Matter* and Moritz Schlick's *Space and Time in Physics Today*, along with another volume entitled *The Principle of Relativity*, whose third edition was to contain the most important of the original essays on general relativity. (See Einstein's letter to Maurice Solovine, April 24, 1920, in *Letters to Solovine*, 21)

HOBBIES

Besides music and reading, Einstein's passion was sailing. For his fiftieth birthday, a group of friends bought him a sailboat, which he sailed in the Havel River at his summer home in Caputh, southwest of Berlin. The boat, a 215-square-foot dinghy outfitted in mahogany, was called *Tümmler* ("dolphin," or "something that glides"; I mistranslated this word in the original edition as "acrobat" based on the verb "tummeln," but "dolphin" or "porpoise" is more likely to be the correct translation). Later, in Princeton, he sailed on Lake Carnegie in his more modest boat, *Tinnef* ("cheaply made" in Yiddish).

HANDEDNESS

Einstein, unlike many physicists and mathematicians, was right-handed. Photos show him pointing with his right finger and holding a pen in his right hand. He also held the bow to his violin in the right hand, though admittedly some left-handed people do this as well. No one ever referred to his being left-handed, so one may presume he was like most of us in that respect.

UNDERSTANDING RELATIVITY THEORY

Einstein denied that he ever made the assertion that only twelve people in the world could understand his theory. He thought that every physicist who studied the theory could readily understand it. (Denial made to reporters upon his arrival in New York City in 1921; see Frank, *Einstein: His Life and Times*, 179.)

*HONORARY DEGREES

Einstein's first honorary degree was awarded by the University of Geneva in 1909. He also received degrees from Harvard University, Lincoln University, Princeton University, State University of New York at Albany, and Yeshiva University in the United States; Buenos Aires University in South America; and the universities of Brussels, Cambridge, Glasgow, Leeds, London, Madrid, Manchester, Oxford, Rostock, the Sorbonne, and Zurich in Europe. He may have been awarded other degrees, perhaps in Japan, though I have not yet come across evidence for them.

EINSTEIN'S AMERICAN CITIZENSHIP

Einstein entered the United States in 1933 with only a visitor's visa. Under U.S. immigration law at the time, permission to become a citizen could be obtained only through an American consul in a foreign country. Einstein therefore chose to go to Bermuda to apply for citizenship in May 1935. The American consul threw a gala dinner in his honor and gave him permission to enter the United States as a permanent resident. Five years later, in 1940, he, Margot Einstein, and Helen Dukas became citizens as they took their oath of allegiance in Trenton, New Jersey. (See Pais, *Einstein Lived Here*, 199; and Frank, *Einstein: His Life and Times*, 293, though Frank gives the wrong year of citizenship.)

AT THE INSTITUTE FOR ADVANCED STUDY, PRINCETON

The Institute's mission has been to attract the world's best scholars and enable them to attend to their work in a peaceful haven of scholarship and collegiality while earning more than fair salaries.

Einstein's starting salary in 1933 was $15,000 per year (quite high for that time, perhaps giving credence to its nickname "Institute for Advanced Salaries"), with a $5,000 yearly pension at retirement (Einstein Archive 29-315). The Institute was provided temporary quarters on the Princeton University campus, in a part of Fine Hall, the old mathematics building, which is now Jones Hall, the home of the East Asian Studies department. In 1940 it was moved to its own campus in a rural part of Princeton. Einstein retired in 1945 but continued to occupy an office at the Institute until his death.

At that time, Abraham Flexner was the director of the Institute and, to Einstein's annoyance, proved to be an overly protective boss. Soon after Einstein's arrival in the United States, for example, President Roosevelt invited Einstein and his wife, Elsa—via the director's office—to the White House. Flexner took it upon himself to decline the invitation without consulting Einstein, citing security reasons. Some time later, Einstein was told about this incident and hastily wrote an apologetic letter to Roosevelt. The invitation was extended once more, and Einstein finally did make it to the White House to meet the president (and he didn't wear socks).

*THE ILSE LETTER

In the spring of 1918, Einstein considered breaking off his engagement to Elsa as he was contemplating marrying her pretty twenty-year-old daughter, Ilse, instead. Ilse had become his secretary at the Kaiser Wilhelm Institute in January 1918. At this time he was also going through the final stages of his divorce from Mileva.

This information became public fairly recently, when volume 8 of *CPAE* was published in 1998. In a letter of May 22, 1918, Ilse confided to Georg Nicolai, a doctor and antiwar crusader, that Einstein had approached her about marriage, and she seemed uncertain about how to handle the delicate situation. Her mother, who had been having an affair with Einstein since 1912, knew about the predicament and, according to the daughter, would be willing to step aside if Ilse's happiness were at stake. But Ilse decided she didn't have the same kind of passion for Einstein as he seemed to have for her—indeed, to her he was more of a father figure (he was thirty-eight years old at the time), and she had no desire to be physically close to him. In the end, Ilse refused the proposal; the following year, Elsa and Einstein were married. (See *CPAE*, Vol. 8, Doc. 545, for the complete letter to Nicolai.)

This incident shows vividly how detached Einstein was from personal things; as he himself admitted, he was better at dealing with the "impersonal" side of life. He was indifferent about a question of great importance to most people—one acceptable woman was just as good as another, and, to his credit, age didn't matter, either; for example, both Elsa and Mileva were older than he, yet Ilse was quite a bit younger.

*THE NOBEL PRIZE

In November 1922, when Einstein was in the Far East at the invitation of *Kaizo* magazine (Japan), he received the news in Shanghai that he had won the Nobel Prize in Physics for 1921 for his "services to theoretical physics and especially for his discovery of the photoelectric effect." There is no record of how he reacted to news of the award, and no mention is made of it in his travel diary (Pais, *Subtle Is the Lord*, 503). Many feel that the Nobel Committee (part of the Royal Swedish Academy of Sciences) snubbed any mention of relativity theory because it had become increasingly controversial, yet the committee was under pressure to give Einstein the prize. To quote Pais, "It was the Academy's bad fortune not to have anyone among its members who could competently evaluate the content of relativity theory in those early years"; and the proposal to award it for the photoelectric effect—also worthy of the prize—was a way out.

Einstein had already been nominated for the Nobel Prize every year from 1910 until 1918, except for 1911 and 1915.

The entire monetary award, amounting to about $32,000 in 1923 money, went to Mileva Marić as part of Einstein's promised divorce settlement with her.

Because Einstein did not return to Berlin from the Far East until the spring of 1923, his Nobel Lecture was delayed. He finally delivered his speech, on the basic ideas and problems of the theory of relativity, in Göteborg, Sweden, in July 1923, in front of an audience of about two thousand.

*MILEVA MARIĆ AS COLLABORATOR

On March 27, 1901, Einstein was deeply engrossed in seeking a job and having difficulty finding one. He wrote to Mileva:

> You are and remain to me a sanctuary that no man can enter; I also know that you love and understand me better than anyone else. I can also assure you that no one here will dare or want to say anything bad about you. How happy and proud I will be when the two of us together will have brought our work on relative motion to a triumphant end! (*CPAE*, Vol. 1, Doc. 94)

The last sentence implies that he and Mileva had been working on "relative motion" together, and the statement caused considerable controversy after it first came to light in the early 1990s. The case for Mileva was championed by her Serbian biographer, Desanka Trbuhović-Gjurić, in her book,

In the Shadow of Albert Einstein; by Serbian physicist Dord Krstić; and by American physician Evan Walker. On the other side were the Einstein scholars, most notably physicist and historian of science John Stachel, the first editor of *The Collected Papers of Albert Einstein*. The debate has not been conclusively resolved, for not enough firsthand evidence exists to prove one side or the other. However, it has been established that most of the case for Mileva has been made through hearsay and fourth-person accounts of Serbian friends and relatives; indeed, Mileva herself never claimed to be a partner in relativity theory. Einstein scholars do not deny that Mileva was intelligent in her own right and probably, as a fellow physics student, played a role as Einstein's helper, supporter, and sounding board for his ideas. She also may have proofread his papers carefully and caught slip-ups and inconsistencies. But to date there is no evidence that she was a creative force behind his work, and to assume so is highly speculative.

A thorough discussion on this topic can be found in Highfield and Carter, *The Private Lives of Albert Einstein*, 108–115.

EINSTEIN'S DEATH AND THE REMOVAL OF HIS BRAIN

Einstein died at Princeton Hospital on April 18, 1955 (see Chronology).

His brain and eyes were removed and preserved, to be saved for future study (see Highfield
and Carter, *The Private Lives of Albert Einstein*, 264ff;
California Monthly, December 1995, 27–28; *Harper's*
magazine, October 1997; *New York Times*, June 18,
1999). The pathologist, Dr. Thomas Harvey, performed an autopsy and, without permission, removed the brain and kept it. Another pathologist,
Dr. Henry Abrams, took the eyes with the permission of the hospital administrator, receiving a letter
of authenticity from Dr. Guy Dean, Einstein's personal physician at the time of his death. It was Einstein's wish that his body be cremated, and his
friends considered the removal of the organs a vio-
lation of his wishes.

After the cremation, the Einstein family learned
about the brain and agreed to let Dr. Harvey keep
it if he did not use it for commercial purposes but
only for scientific study. He then gave at least three
parts of the organ to other scientists, but until recently only Professor Marian Diamond of the
University of California at Berkeley had made a
scientific contribution. In 1985, in an article in *Experimental Neurology*, she reported that Einstein's
brain had an above-average number of glial cells
(which nourish neurons) in those areas of the left
hemisphere that are thought to control mathematical and linguistic skills. An enlarged image of Einstein's glial cells has been on view at the Lawrence
Hall of Science in Berkeley.

Since then, Sandra Witelson, a neuroscientist at McMaster University in Ontario, Canada, published some research results on the brain in June 1999, in the British medical journal *Lancet*. Witelson's group conducted the only study of the overall anatomy of Einstein's brain after Dr. Harvey offered to give them a section of it in 1996. The researchers compared Einstein's brain with the preserved brains of thirty-five men and fifty-six women known to have normal intelligence when they died. They discovered that in Einstein's case the part of the brain thought to be related to mathematical reasoning—the inferior parietal lobe—was 15 percent wider than normal on both sides. Furthermore, they found that the Sylvian fissure, the groove that normally runs from the front of the brain to the back, did not extend all the way in Einstein's case. Witelson theorizes that this latter feature may be the key to Einstein's intelligence, because the absence of a full groove may have allowed more neurons in this area to establish connections among one another and work together more easily. Other parts of Einstein's brain appeared to be a bit smaller than average, putting overall brain size and weight within a normal range.

It may be of interest to some that a few strands of Einstein's hair also remain in the hands of a collector, John Reznikoff of University Archives in Westport, Conn. (www.universityarchives.com).

Einstein's body was taken to the Mather-Hodge Funeral Home in Princeton (which still exists today) and cremated in Trenton on the day of his death. His ashes were scattered in an undisclosed place by two friends, Otto Nathan and Paul Oppenheim. The last person to see Einstein alive was nurse Alberta Rozsel, who reported, "He gave two breaths and expired" (*New York Times*, April 19, 1955). His stepdaughter Margot described the last hours this way in a letter the same month to family friend Hedwig Born: "He . . . waited for his end as for an impending natural event. He faced death quietly and modestly, and was fearless as he had been in life. He left this world without sentimentality and without regrets" (see Born, *Born-Einstein Letters*, 234). Helen Dukas also recorded an account of Einstein's last days (Einstein Archive 39-071).

A memorial concert took place at McCarter Theater in Princeton later that year, on December 17, 1955, and featured the following program: R. Casadesus, piano, and the Princeton University Orchestra performing Mozart's Coronation Concerto (concerto for piano and orchestra in D Major) and Bach's Sonatina from Cantata no. 106 "Actus Tragicus." Also played were Haydn's Symphony no. 104 in D Major and Corelli's Concerto Grosso no. 8 ("Christmas"—presumably because of the holiday season).

MISCELLANEOUS PERSONAL INFORMATION

Einstein did not learn to speak until he was two and a half (according to the biography of him by his sister, Maja) or three (according to other sources). It has been suggested that this lateness was the origin of his thinking in visual terms (i.e., his "thought experiments").

Einstein was a better than average student in school. His highest marks were in mathematics, physics, and music. His lowest were in French and Italian (*CPAE*, Vol. 1, Docs. 8 and 10).

Einstein's Swiss military service book shows the following result of a health examination that deemed him unfit for military service at the age of 22 (March 13, 1901):

Body height, 171.5 cm (5 ft, 7.6 in)

Chest circumference, 87 cm (34.8 in)

Upper arm, 28 cm (11.2 in)

Diseases or defects: varicose veins, flat feet, and excessive foot perspiration

(see *CPAE*, Vol. 1, Doc. 91). According to Helen Dukas, Einstein was required to pay a tax until 1940 for not serving in the Swiss military; he kept a *Dienstbuch* (service book) from the military that showed yearly entries of tax payments.

In 1920, Einstein asked Princeton University for a $15,000 honorarium for two months of lecturing, three lectures a week (Einstein Archive 36-241). However, the Princeton lectures, delivered in 1921 and published in 1922, were cut down to four lectures entitled "The Meaning of Relativity," for which he received a much smaller fee.

Einstein generally wrote letters that were short and to the point, especially in his later years. The longest handwritten letter I came across while working in the archive was ten pages long; it was written to physicist H. A. Lorentz on January 23, 1915 (Einstein Archive 16-436).

Einstein claimed he got his hairstyle "through negligence."

Einstein and his family were animal lovers. In Princeton they kept a dog named Chico and a cat named Tiger.

Einstein did not seem to like to have his manuscripts reviewed by others. In the summer of 1936 he submitted a paper entitled "On Gravitational Waves" to the *Physical Review*, and a referee returned it with ten pages of comments. Insulted, Einstein, along with coauthor Nathan Rosen, withdrew the paper so that they could publish it elsewhere. They did so the following year, in the *Journal of the Franklin Institute*. Einstein claimed the *Physical Review* had no right to show the paper to reviewers before publication, as was (and is) the American custom. (See an excerpt from his letter to the editor of the *Physical Review*, July 27, 1936, in the section "On Science." Einstein Archive 19-087.)

Grete Markstein, a talented actress, claimed to be Einstein's daughter until the day she died. To disprove her claim, Einstein, on Helen Dukas's initiative, had her birth records checked, and they proved otherwise (for example, she was only thirteen years younger than he). She died in 1947. (From notes of a conversation with Helen Dukas.) Both Einstein and his friend János Plesch wrote humorous poems about Mrs. Markstein (too difficult to translate poetically from the German; Einstein Archive 31-540 and 31-541).

Einstein never allowed his name to be used for commercial advertising, though he received some curious requests, for example, from a company manufacturing "hair restorers" and soap; also from a maker of pens. If he as much as showed enthusiasm for a product, word would get around and he would be approached to endorse and promote it. Today his estate employs a California-based advertising agency, the Roger Richman Agency, to protect commercial use of his name with a trademark.

By middle and old age, Einstein came to harbor bitter feelings against the opposite sex, calling marriage incompatible with human nature: marriage makes people treat each other as property and one can no longer operate as a free human being. All the men in Einstein's nuclear family preferred older women as partners: both of Einstein's wives were at least three years older than he; son Hans Albert's first wife was nine years older, his second wife two years older; son Eduard had an older woman friend but never married.

Einstein snored "unbelievably loudly," according to Elsa, so they kept separate bedrooms at home and when traveling. Elsa was not allowed to enter his study—he demanded complete privacy there. To Elsa: "Speak of you *or* me, but never of 'us.'" He rarely used "we" for them as a couple, either, preferring to speak only for himself.

Einstein in September 1945 with Leo Szilard, who had encouraged him to write a letter to President Roosevelt warning him about the possibility that nuclear chain reactions could lead to the building of atom bombs. Szilard, along with Enrico Fermi, achieved the world's first sustained nuclear fission reaction. (March of Time/National Archives)

I had always wondered what a Federal Bureau of Investigation file looks like. I had imagined clean, carefully organized papers collected by sophisticated spymasters who led lives of intrigue and danger and passed messages to one another by way of public phone booths and clever code words. Indeed, when I was a child I had often fancied that I would become one of them myself.

But after I finished taking a peek into the Einstein file, I was glad I had chosen a different career path. What I saw left me incredulous that such clandestine invasions into a person's life could take place in a "free" America. To me it smacked of Communism and the KGB, yet ironically the purpose of the file was to help check the perceived Communist threat in the United States.

Twenty-five years after Einstein's death, the FBI revealed that J. Edgar Hoover had been keeping a secret file on Einstein's political activities at least since 1940. The reasons were that Einstein had been seen alongside Communists attending pacifist meetings and he had supported the republican cause in Spain during the Spanish Civil War. (See A. Summers, *Official and Confidential: The Secret Life of J. Edgar Hoover*. London: Victor Golancz, 1933.)

He never received the security clearance that was necessary to work on the Manhattan Project's development of the atom bomb. (See also Einstein's second letter to President Roosevelt, below.) Einstein had not been aware of this intrusion into his privacy, which lasted until his death. There are 1,427 pages of material, according to the FBI's form letter to me, and I could have it all at a cost of $132.70 under the Freedom of Information Act. It wasn't a bad deal, but I decided to scrutinize the file in the Einstein duplicate archive at Boston University instead.

The file was, to my eyes, an eyesore. It consists of a dossier filled with a muddle of memos from FBI agents to one another about what the "subject" or "captioned individual" had been up to. There are notes and observations about meetings Einstein attended and organizations that listed him as a member. Most documents in fact indicate that he was not a Communist or subversive, yet the file remained active, probably because some of the organizations he allegedly joined or supported were considered subversive. Indeed, in a letter of July 26, 1940, about two months before Einstein became an American citizen, he was denied security clearance by Brigadier General George Strong.

In all of the memos, certain words and passages are blocked out, perhaps to protect innocent people who are still alive. An assortment of rubber-stamped locutions shout out at you, the loudest of which warn you that this page is "SECRET" or "CON-FIDENTIAL," though these words were crossed out when the file was declassified. In addition, there are dozens of newspaper clippings covering Einstein's activities, probably secured through an internal clipping service. (See also R. A. Schwartz, "Einstein and the War Department," *Isis* 80 [1989], 281–284.)

SOME SAMPLES

A memo discusses the "Committee of 1,000," the plan of Harlow Shapley of Harvard to recruit one thousand prominent Americans, including Einstein, in a drive to abolish the House Un-American Activities Committee (HUAC).

A memo discusses *Counterattack*, a weekly newsletter published by the American Business Consultants Inc. of New York. It ran an article about a meeting of the American Committee of Jewish Writers, Artists, and Scientists, which it claimed was a Communist-front organization, and maintained that Einstein had allowed himself to be "roped" into it.

A memo claims Einstein made the following statement in December 1947: "I came to America because of the great, great freedom which I heard existed in this country. I made a mistake in selecting America as a land of freedom, a mistake I cannot repair in the balance of my life."

Files dated February 13 and 15, 1950, state that Einstein had no formal security clearance from the Atomic Energy Commission or the Manhattan Engineer District. This confirms that Einstein could not have worked on the atom bomb.

Oddly, one of the memos continues that "the Bureau files fail to reflect that any investigation has ever been conducted on Professor Einstein for any purpose whatsoever."

The memo then states that Einstein was affiliated in "some way or other" with at least thirty-three organizations that were cited by the Attorney General, HUAC, or California HUAC as Communist organizations. He was also affiliated with fifty organizations that were not so cited. "He is principally a pacifist and could be considered a liberal thinker as indicated by his connections with the various organizations indicated above."

Mention is made that Einstein is sympathetic to Russian scientists and sympathizes with the Soviet Union. Supplementary handwritten notes mention that Einstein's son is in Russia.

Another memo quotes an informant as saying that Elsa Einstein was "scared to death" over the fact that son Hans Albert was in the Soviet Union around 1944 and might be held hostage there to force Einstein into some particular action. (No matter that Elsa had died in 1936, and that Hans Albert was her *step*son.)

A memo of March 13, 1950, claims that Einstein's office in Berlin had been used as a telegram address for Soviet Comintern agents and other Soviet "apparati" in the early 1930s, before he left Germany, and that he had hired a group of typists and secretaries who were Soviet sympathizers. (In fact, Helen Dukas was his only secretary at the time.) The memo states that one of Einstein's secretaries turned over the telegrams ("conspirative correspondence") to a special "apparat man" (a Soviet courier) who also picked up telegrams at other designated addresses. Einstein's address was considered a useful cover for a letter drop because he received telegrams from all over the world. The telegrams were in code, and it was assumed Einstein did not know their contents.

A memo of September 14, 1950, states that "this naturalized person, notwithstanding his worldwide reputation as a scientist, may properly be investigated for possible revocation of naturalization. . . . It appears that appropriate investigation for that purpose is warranted." This investigation was warranted because of the alleged Soviet use of Einstein's address in Berlin.

A memo of October 23, 1950, from J. Edgar Hoover himself requests that Helen Dukas be investigated for "past activity on behalf of the Soviet Union" (i.e., while serving as Einstein's secretary in Berlin).

Later memos fault Einstein for allowing Paul Robeson to deliver Einstein's letter to President Harry Truman stating his opposition to lynching. Robeson was chairman of the American Crusade to End Lynching, an alleged Communist-front organization. Another memo lists "Indicators of Einstein's Sympathy with the Communist Party."

The complete file is now available on the Internet at www.fbi.gov.

THE FAMOUS LETTER TO
PRESIDENT FRANKLIN D. ROOSEVELT

Peconic, Long Island,
August 2nd, 1939

Sir:

Some recent work by E. Fermi and L. Szilard, which has been communicated to me in manuscript, leads me to expect that the element uranium may be turned into a new and important source of energy in the immediate future. Certain aspects of the situation which has arisen seem to call for watchfulness and, if necessary, quick action on the part of the Administration. I believe therefore that it is my duty to bring to your attention the following facts and recommendations:

In the course of the last four months it has been made probable—through the work of Joliot in France as well as Fermi and Szilard in America—that it may become possible to set up a nuclear chain reaction in a large mass of uranium, by which vast amounts of power and large quantities of new radium-like elements would be generated. Now it appears almost certain that this could be achieved in the immediate future.

This new phenomenon would also lead to the construction of bombs, and it is conceivable—though much less certain—that extremely powerful bombs of a new type may thus be constructed. A single bomb of this type, carried by boat and exploded in a port, might very well destroy the whole port together with some of the surrounding territory. However, such bombs might very well prove to be too heavy for transportation by air.

The United States has only very poor ores of uranium in moderate quantities. There is some good ore in Canada and the former Czechoslovakia, while the most important source of uranium is [the] Belgian Congo.

In view of this situation you may think it desirable to have some permanent contact maintained between the Administration and the group of physicists working on chain reactions in America. One possible way of achieving this might be for you to entrust with this task a person who has your confidence and who could perhaps serve in an inofficial [*sic*] capacity. His task might comprise the following:

a) to approach Government Departments, keep them informed of the further development, and put forward recommendations for Government action, giving particular attention to the problem of securing a supply of uranium ore for the United States;

b) to speed up the experimental work, which is at present being carried on within the limits of the budgets of University laboratories, by providing funds, if such funds be required, through his contacts with private persons who are willing to make contributions for this cause, and perhaps also by obtaining the co-operation of industrial laboratories which have the necessary equipment.

I understand that Germany has actually stopped the sale of uranium from the Czechoslovakian mines which she has taken over. That she should have taken such early action might perhaps be understood on the ground that the son of the German Under-Secretary of State, von Weizsäcker, is attached to the Kaiser-Wilhelm-Institut in Berlin where some of the American work on uranium is now being repeated.

> *Yours very truly,*
> *Albert Einstein*

Roosevelt replied, in part, on October 19, 1939, as follows: "I have found this letter of such import that I have convened a board consisting of the head of the Bureau of Standards and chosen representatives of the Army and Navy to thoroughly investigate the possibilities of your suggestion regarding the element of uranium." (See Rosenkranz, *Albert through the Looking-Glass*, 66–67. Einstein Archive 33-088. Einstein's letter was bought at auction by the Forbes family.)

The new board met only two days later, on October 21, with Enrico Fermi, Leo Szilard, Edward Teller, and Eugene Wigner serving as experts on nuclear fission.

A lesser-known letter was sent to Roosevelt five and a half years later, as Einstein came to fear the possible misuse of uranium.

> *March 25, 1945*

Sir:

I am writing you to introduce Dr. L. Szilard, who proposes to submit to you certain considerations and recommendations. Unusual circumstances which I shall describe further below induce me to take this action in spite of the fact that I do not know the substance of the considerations and recommendations which Dr. Szilard proposes to submit to you.

In the summer of 1939 Dr. Szilard put before me his views concerning the potential importance of uranium for national defense. He was greatly disturbed by the potentialities involved and anxious that the United States Government be advised of them as soon as possible. Dr. Szilard, who is one of the discoverers of the neutron emission of uranium on which all present work on uranium is based, described to me a specific system which he devised and which he thought would make it possible to set up a chain reaction in unseparated uranium in the immediate future. Having known him for over twenty years both from his scientific work and personally, I have much confidence in his judgment, and it was on the basis of his judgment as well as my own that I took the liberty to approach you in connection with this subject. You responded to my letter dated August 2, 1939 by the appointment of a committee under the chairmanship of Dr. Briggs and thus started the Government's activity in this field.

The terms of secrecy under which Dr. Szilard is working at present do not permit him to give me information about his work; however, I understand that he now is greatly concerned about the lack of adequate contact between scientists who are doing this work and those members of your Cabinet who are responsible for formulating policy. In the circumstances, I consider it my duty to give Dr. Szilard this introduction and I wish to express the hope that you will be able to give his presentation of the case your personal attention.

Very truly yours,
A. Einstein

Roosevelt died April 12, 1945, of a cerebral hemorrhage. It is believed he never saw this letter.

Bibliography

Aichenburg, P., and R. Sexl. *Albert Einstein*. Braunschweig: Vieweg, 1979.

Bernstein, Jeremy. *Einstein*. New York: Penguin, 1978.

Born, Max, ed. *Einstein-Born Briefwechsel, 1916–1955*. Munich: Nymphenbürger, 1969.

———. *The Born-Einstein Letters*. Trans. Irene Born. New York: Walker, 1971.

Cassidy, David. *Einstein and Our World*. Atlantic Highlands, N.J.: Humanities Press, 1995.

Clark, Ronald W. *Einstein: The Life and Times*. New York: Crowell, 1971.

Cline, Barbara Lovett. *Men Who Made a New Physics*. Chicago: University of Chicago Press, 1987.

CPAE. See Stachel et al. for Vols. 1 and 2; Klein et al. for Vol. 5; Kox et al. for Vol. 6; Janssen et al. for Vol. 7; Schulmann et al. for Vols. 8 and 9.

Cuny, Hilaire. *Albert Einstein: The Man and His Times*. London, 1963.

Dukas, Helen, and Banesh Hoffmann. *Albert Einstein, the Human Side*. Princeton, N.J.: Princeton University Press, 1979.

Dürrenmatt, Friedrich. *Albert Einstein: Ein Vortrag*. Zurich: Diogenes, 1979.

Dyson, Freeman. "Writing a Foreword for Alice Calaprice's New Einstein Book." *Princeton University Library Chronicle* 57, no. 3 (Spring 1996), 491–502.

Ehlers, Anita. *Liebes Hertz!* Berlin: Birkhäuser, 1994.

Einstein, Albert. *Cosmic Religion with Other Opinions and Aphorisms*. New York: Covici-Friede, 1931.

———. *About Zionism*. Trans. L. Simon. New York: Macmillan, 1931.

Einstein, Albert. *The World as I See It*. Abridged ed. New York: Philosophical Library, distributed by Citadel Press. Orig. in Leach, *Living Philosophies*, 1931.

———. *The Origins of the Theory of Relativity*. Glasgow: Jackson, Wylie, 1933.

———. *Essays in Science*. Trans. Alan Harris. New York: Philosophical Library, 1934.

———. *Mein Weltbild*. Amsterdam: Querido Verlag, 1934. Paperback ed., Berlin: Ullstein, 1993.

———. *Out of My Later Years*. Paperback ed. New York: Wisdom Library of the Philosophical Library, 1950. (Other editions exist as well; page numbers refer to this edition.)

———. *Ideas and Opinions*. Trans. Sonja Bargmann. New York: Crown, 1954. (Other editions exist as well; page numbers refer to this edition.)

———. *Albert Einstein/Mileva Marić: The Love Letters*. Ed. Jürgen Renn and Robert Schulmann. Trans. Shawn Smith. Princeton, N.J.: Princeton University Press, 1992.

———. *Einstein on Humanism*. New York: Carol Publishing, 1993.

———. *Letters to Solovine, 1906–1955*. Trans. from the French by Wade Baskin, with facsimile letters in German. New York: Carol Publishing, 1993.

Einstein, Albert, and Sigmund Freud. *Why War?* Paris: Institute for Intellectual Cooperation, League of Nations, 1933.

Einstein, Albert, and Leopold Infeld. *The Evolution of Physics*. New York: Simon and Schuster, 1938.

Einstein: A Portrait. Corte Madera, Calif.: Pomegranate Artbooks, 1984.

Fölsing, Albrecht. *Albert Einstein*. Trans. Ewald Osers. New York: Viking, 1997.

Frank, Philipp. *Einstein: His Life and Times*. New York: Knopf, 1947, 1953.

———. *Einstein: Sein Leben und seine Zeit*. Braunschweig: Vieweg, 1979.

French, A. P., ed. *Einstein: A Centenary Volume*. Cambridge, Mass.: Harvard University Press, 1979.

Grüning, Michael. *Ein Haus für Albert Einstein*. Berlin: Verlag der Nation, 1990.

Hadamard, Jacques. *An Essay on the Psychology of Invention in the Mathematical Field*. Princeton, N.J.: Princeton University Press, 1945.

Hermann, Armin. *Albert Einstein*. Munich: Piper, 1994.

Highfield, Roger, and Paul Carter. *The Private Lives of Albert Einstein*. London: Faber and Faber, 1993.

Hoffmann, Banesh. *Albert Einstein: Creator and Rebel*. New York: Viking, 1972.

———. "Einstein and Zionism." In *General Relativity and Gravitation*, ed. G. Shaviv and J. Rosen. New York: Wiley, 1975.

Holton, Gerald. *The Advancement of Science and Its Burdens*. New York: Cambridge University Press, 1986.

Holton, Gerald, and Yehuda Elkana, eds. *Albert Einstein: Historical and Cultural Perspectives. The Centennial Symposium in Jerusalem*. Princeton, N.J.: Princeton University Press, 1982.

Infeld, Leopold. *The Quest: The Evolution of a Scientist*. New York: Doubleday, 1941.

———. *Albert Einstein*. Rev. ed. New York: Charles Scribner's Sons, 1950.

Jammer, Max. *Einstein and Religion*. Princeton, N.J.: Princeton University Press, 1999.

Janssen, Michel, Robert Schulmann, and Christoph Lehner, eds. *The Collected Papers of Albert Einstein*, Vol. 7, *The Berlin Years: Writings, 1918–1921*. Princeton, N.J.: Princeton University Press, forthcoming. (Trans. Alfred Engel, forthcoming.)

Kaller's Autographs Catalog. "Jewish Visionaries," 1997. Kaller's Antiques and Autographs, at Macy's, 37th St., New York City.

Klein, Martin, A. J. Kox, and Robert Schulmann, eds. *The Collected Papers of Albert Einstein*, Vol. 5, *The Swiss Years:*

Correspondence, 1902–1914. Princeton, N.J.: Princeton University Press, 1993. (Trans. Anna Beck, 1995.)

Kox, A. J., Martin J. Klein, and Robert Schulmann, eds. *The Collected Papers of Albert Einstein*, Vol. 6, *The Berlin Years: Writings, 1914–1917*. Princeton, N.J.: Princeton University Press, 1996. (Trans. Alfred Engel, 1996.)

Leach, Henry J., ed. *Living Philosophies: A Series of Intimate Credos*. New York: Simon and Schuster, 1931.

Michelmore, P. *Einstein: Profile of the Man*. New York: Dodd, 1962.

Moszkowski, Alexander. *Conversations with Einstein*. Trans. Henry L. Brose. New York: Horizon Press, 1970. (Conversations took place in 1920, trans. 1921, published in English in 1970.)

Nathan, Otto, and Heinz Norden, eds. *Einstein on Peace*. New York: Simon and Schuster, 1960.

Pais, Abraham. *Subtle Is the Lord: The Science and the Life of Albert Einstein*. Oxford: Oxford University Press, 1982.

———. *Einstein Lived Here*. Oxford: Oxford University Press, 1994.

———. *A Tale of Two Continents*. Princeton, N.J.: Princeton University Press, 1997.

Planck, Max. *Where Is Science Going?* New York: Norton, 1932.

Regis, Ed. *Who Got Einstein's Office?* Reading, Mass.: Addison-Wesley, 1987.

Reines, Frederick, ed. *Cosmology, Fusion and Other Matters: A Memorial to George Gamow*. Boulder: University Press of Colorado, 1972.

Reiser, Anton. *Albert Einstein: A Biographical Portrait*. New York: Boni, 1930.

Rosenkranz, Ze'ev. *Albert through the Looking-Glass: The Personal Papers of Albert Einstein*. Jerusalem: The Jewish National and University Library, 1998.

Rosenthal-Schneider, Ilse. *Reality and Scientific Truth*. Detroit: Wayne State University Press, 1980.

Ryan, Dennis P., ed. *Einstein and the Humanities*. New York: Greenwood Press, 1987.

Sayen, Jamie. *Einstein in America*. New York: Crown, 1985.

Schilpp, Paul, ed. *Albert Einstein: Philosopher-Scientist*. Evanston, Ill.: Library of Living Philosophers, 1949.

———, ed. and trans. *Albert Einstein: Autobiographical Notes*. Paperback ed. La Salle, Ill.: Open Court, 1979. (These *Notes* are also contained in the preceding volume.)

Schulmann, Robert. "Einstein Rediscovers Judaism." Unpublished manuscript, 1999.

Schulmann, Robert, A. J. Kox., Michel Janssen, and József Illy, eds. *The Collected Papers of Albert Einstein*, Vol. 8, Parts A and B, *The Berlin Years: Correspondence, 1914–1918*. Princeton, N.J.: Princeton University Press, 1998. (Trans. Ann Lehar, 1998.)

Schulmann, Robert, Michel Janssen, and József Illy, eds. *The Collected Papers of Albert Einstein*, Vol. 9, *The Berlin Years: Correspondence, 1919–1921*. Princeton, N.J.: Princeton University Press, forthcoming. (Trans. Ann Lehar, forthcoming.)

Seelig, Carl, ed. *Helle Zeit, dunkle Zeit: In Memorium Albert Einstein*. Zurich: Europa Verlag, 1956.

Stachel, John. "Einstein's Jewish Identity." Unpublished manuscript, 1989.

Stachel, John, ed., with the assistance of Trevor Lipscombe, Alice Calaprice, and Sam Elworthy. *Einstein's Miraculous Year: Five Papers That Changed the Face of Physics*. Princeton, N.J.: Princeton University Press, 1998.

Stachel, John, et al., eds. *The Collected Papers of Albert Einstein*, Vol. 1, *The Early Years: 1879–1902*. Princeton, N.J.: Princeton University Press, 1987. (Trans. Anna Beck, 1987.)

———. *The Collected Papers of Albert Einstein*, Vol. 2, *The Swiss Years: Writings, 1900–1909*. Princeton, N.J.: Princeton University Press, 1989. (Trans. Anna Beck, 1989.)

Stern, Fritz. *Einstein's German World*. Princeton, N.J.: Princeton University Press, 1999.

Vallentin, Antonina. *Das Drama Albert Einsteins*. Stuttgart: Günther Verlag, 1955.

Whitrow, G. J. *Einstein: The Man and His Achievement*. New York: Dover, 1967.

Index of Key Words

Subject Index